Springer Undergraduate Mathematics Series

Springer
London
Berlin
Heidelberg
New York
Barcelona
Hong Kong
Milan
Paris
Singapore
Tokyo

Other books in this series

A First Course in Discrete Mathematics *I. Anderson*
Analytic Methods for Partial Differential Equations *G. Evans, J. Blackledge, P. Yardley*
Applied Geometry for Computer Graphics and CAD *D. Marsh*
Basic Linear Algebra *T.S. Blyth and E.F. Robertson*
Basic Stochastic Processes *Z. Brzeźniak and T. Zastawniak*
Elementary Differential Geometry *A. Pressley*
Elementary Number Theory *G.A. Jones and J.M. Jones*
Elements of Logic via Numbers and Sets *D.L. Johnson*
Groups, Rings and Fields *D.A.R. Wallace*
Hyperbolic Geometry *J.W. Anderson*
Information and Coding Theory *G.A. Jones and J.M. Jones*
Introduction to Laplace Transforms and Fourier Series *P.P.G. Dyke*
Introduction to Ring Theory *P.M. Cohn*
Introductory Mathematics: Algebra and Analysis *G. Smith*
Introductory Mathematics: Applications and Methods *G.S. Marshall*
Linear Functional Analysis *B.P. Rynne and M.A. Youngson*
Measure, Integral and Probability *M. Capiński and E. Kopp*
Multivariate Calculus and Geometry *S. Dineen*
Numerical Methods for Partial Differential Equations *G. Evans, J. Blackledge, P. Yardley*
Sets, Logic and Categories *P. Cameron*
Topics in Group Theory *G.C. Smith and O. Tabachnikova*
Topologies and Uniformities *I.M. James*
Vector Calculus *P.C. Matthews*

P.P.G. Dyke

An Introduction to Laplace Transforms and Fourier Series

With 51 Figures

Springer

Philip P.G. Dyke, BSc, PhD
Professor of Applied Mathematics, University of Plymouth, Drake Circus, Plymouth,
Devon, PL4 8AA, UK

Cover illustration elements reproduced by kind permission of:
Aptech Systems, Inc., Publishers of the GAUSS Mathematical and Statistical System, 23804 S.E. Kent-Kangley Road, Maple Valley, WA 98038,
USA. Tel: (206) 432 - 7855 Fax (206) 432 - 7832 email: info@aptech.com URL: www.aptech.com
American Statistical Association: Chance Vol 8 No 1, 1995 article by KS and KW Heiner 'Tree Rings of the Northern Shawangunks' page 32 fig 2
Springer-Verlag: Mathematica in Education and Research Vol 4 Issue 3 1995 article by Roman E Maeder, Beatrice Amrhein and Oliver Gloor
'Illustrated Mathematics: Visualization of Mathematical Objects' page 9 fig 11, originally published as a CD ROM 'Illustrated Mathematics' by
TELOS: ISBN 0-387-14222-3, German edition by Birkhauser: ISBN 3-7643-5100-4.
Mathematica in Education and Research Vol 4 Issue 3 1995 article by Richard J Gaylord and Kazume Nishidate 'Traffic Engineering with Cellular
Automata' page 35 fig 2. Mathematica in Education and Research Vol 5 Issue 2 1996 article by Michael Trott 'The Implicitization of a Trefoil
Knot' page 14.
Mathematica in Education and Research Vol 5 Issue 2 1996 article by Lee de Cola 'Coins, Trees, Bars and Bells: Simulation of the Binomial Pro-
cess page 19 fig 3. Mathematica in Education and Research Vol 5 Issue 2 1996 article by Richard Gaylord and Kazume Nishidate 'Contagious
Spreading' page 33 fig 1. Mathematica in Education and Research Vol 5 Issue 2 1996 article by Joe Buhler and Stan Wagon 'Secrets of the
Madelung Constant' page 50 fig 1.

British Library Cataloguing in Publication Data
Dyke, P.P.G.
   An introduction to Laplace transforms and Fourier series. -
   (Springer undergraduate mathematics series)
   1. Fourier series 2. Laplace transformation 3. Fourier
   transformations 4. Fourier series – Problems, exercises,
   etc. 5. Laplace transformations – Problems, exercises, etc.
   6. Fourier transformations – Problems, exercises, etc.
   I. Title
   515.7'23
ISBN 1852330155

Library of Congress Cataloging-in-Publication Data
Dyke, P.P.G.
   An introduction to Laplace transforms and Fourier series. / P.P.G. Dyke
      p.   cm. -- (Springer undergraduate mathematics series)
   Includes index.
   ISBN 1-85233-015-5 (alk. paper)
   1. Laplace transformation. 2. Fourier series. I. Title. II. Series.
QA432.D94   1999   98-47927
515'.723—dc21   CIP

Springer Undergraduate Mathematics Series ISSN 1615-2085
ISBN 1-85233-015-5 Springer-Verlag London Berlin Heidelberg
Springer-Verlag is a part of Springer Science+Business Media
springeronline.com

Typesetting: Camera ready by the author and Michael Mackey
Printed and bound at the Athenæum Press Ltd., Gateshead, Tyne & Wear
12/3830-5432 Printed on acid-free paper SPIN 10980033

To Ottilie

Preface

This book has been primarily written for the student of mathematics who is in the second year or the early part of the third year of an undergraduate course. It will also be very useful for students of engineering and the physical sciences for whom Laplace Transforms continue to be an extremely useful tool. The book demands no more than an elementary knowledge of calculus and linear algebra of the type found in many first year mathematics modules for applied subjects. For mathematics majors and specialists, it is not the mathematics that will be challenging but the applications to the real world. The author is in the privileged position of having spent ten or so years outside mathematics in an engineering environment where the Laplace Transform is used in anger to solve real problems, as well as spending rather more years within mathematics where accuracy and logic are of primary importance. This book is written unashamedly from the point of view of the applied mathematician.

The Laplace Transform has a rather strange place in mathematics. There is no doubt that it is a topic worthy of study by applied mathematicians who have one eye on the wealth of applications; indeed it is often called Operational Calculus. However, because it can be thought of as specialist, it is often absent from the core of mathematics degrees, turning up as a topic in the second half of the second year when it comes in handy as a tool for solving certain breeds of differential equation. On the other hand, students of engineering (particularly the electrical and control variety) often meet Laplace Transforms early in the first year and use them to solve engineering problems. It is for this kind of application that software packages (MATLAB©, for example) have been developed. These students are not expected to understand the theoretical basis of Laplace Transforms. What I have attempted here is a mathematical look at the Laplace Transform that demands no more of the reader than a knowledge of elementary calculus. The Laplace Transform is seen in its typical guise as a handy tool for solving practical mathematical problems but, in addition, it is also seen as a particularly good vehicle for exhibiting fundamental ideas such as a mapping, linearity, an operator, a kernel and an image. These basic principals are covered

vii

in the first three chapters of the book. Alongside the Laplace Transform, we develop the notion of Fourier series from first principals. Again no more than a working knowledge of trigonometry and elementary calculus is required from the student. Fourier series can be introduced via linear spaces, and exhibit properties such as orthogonality, linear independence and completeness which are so central to much of mathematics. This pure mathematics would be out of place in a text such as this, but Appendix C contains much of the background for those interested. In Chapter 4 Fourier series are introduced with an eye on the practical applications. Nevertheless it is still useful for the student to have encountered the notion of a vector space before tackling this chapter. Chapter 5 uses both Laplace Transforms and Fourier series to solve partial differential equations. In Chapter 6, Fourier Transforms are discussed in their own right, and the link between these, Laplace Transforms and Fourier series is established. Finally, complex variable methods are introduced and used in the last chapter. Enough basic complex variable theory to understand the inversion of Laplace Transforms is given here, but in order for Chapter 7 to be fully appreciated, the student will already need to have a working knowledge of complex variable theory before embarking on it. There are plenty of sophisticated software packages around these days, many of which will carry out Laplace Transform integrals, the inverse, Fourier series and Fourier Transforms. In solving real-life problems, the student will of course use one or more of these. However this text introduces the basics; as necessary as a knowledge of arithmetic is to the proper use of a calculator.

At every age there are complaints from teachers that students in some respects fall short of the calibre once attained. In this present era, those who teach mathematics in higher education complain long and hard about the lack of stamina amongst today's students. If a problem does not come out in a few lines, the majority give up. I suppose the main cause of this is the computer/video age in which we live, in which amazing eye catching images are available at the touch of a button. However, another contributory factor must be the decrease in the time devoted to algebraic manipulation, manipulating fractions etc. in mathematics in the 11–16 age range. Fortunately, the impact of this on the teaching of Laplace Transforms and Fourier series is perhaps less than its impact in other areas of mathematics. (One thinks of mechanics and differential equations as areas where it will be greater.) Having said all this, the student is certainly encouraged to make use of good computer algebra packages (e.g. MAPLE©, MATHEMATICA©, DERIVE©, MACSYMA©) where appropriate. Of course, it is dangerous to rely totally on such software in much the same way as the existence of a good spell-checker is no excuse for giving up the knowledge of being able to spell, but a good computer algebra package can facilitate factorisation, evaluation of expressions, performing long winded but otherwise routine calculus and algebra. The proviso is always that students must *understand* what they are doing before using packages as even modern day computers can still be extraordinarily dumb!

In writing this book, the author has made use of many previous works on the subject as well as unpublished lecture notes and examples. It is very difficult to know the precise source of examples especially when one has taught the material to students for some years, but the major sources can be found in the bibliography. I thank an anonymous referee for making many helpful suggestions. It is also a great pleasure to thank my daughter Ottilie whose familiarity and expertise with certain software was much appreciated and it is she who has produced many of the diagrams. The text itself has been produced using LATEX.

P P G Dyke
Professor of Applied Mathematics
University of Plymouth
January 1999

Contents

1
The Laplace Transform

1.1 Introduction

As a discipline, mathematics encompasses a vast range of subjects. In pure mathematics an important concept is the idea of an axiomatic system whereby axioms are proposed and theorems are proved by invoking these axioms logically. These activities are often of little interest to the applied mathematician to whom the pure mathematics of algebraic structures will seem like tinkering with axioms for hours in order to prove the obvious. To the engineer, this kind of pure mathematics is even more of an anathema. The value of knowing about such structures lies in the ability to generalise the "obvious" to other areas. These generalisations are notoriously unpredictable and are often very surprising. Indeed, many say that there is no such thing as non-applicable mathematics, just mathematics whose application has yet to be found.

The Laplace Transform expresses the conflict between pure and applied mathematics splendidly. There is a temptation to begin a book such as this on linear algebra outlining the theorems and properties of normed spaces. This would indeed provide a sound basis for future results. However most applied mathematicians and all engineers would probably turn off. On the other hand, engineering texts present the Laplace Transform as a toolkit of results with little attention being paid to the underlying mathematical structure, regions of validity or restrictions. What has been decided here is to give a brief introduction to the underlying pure mathematical structures, enough it is hoped for the pure mathematician to appreciate what kind of creature the Laplace Transform is, whilst emphasising applications and giving plenty of examples. The point of view from which this book is written is therefore definitely that of the applied mathematician. However, pure mathematical asides, some of which can be quite

1

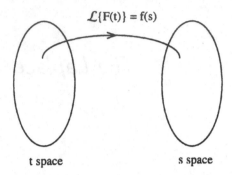

Figure 1.1: The Laplace Transform as a mapping

extensive, will occur. It remains the view of this author that Laplace Transforms only come alive when they are used to solve real problems. Those who strongly disagree with this will find pure mathematics textbooks on integral transforms much more to their liking.

The main area of pure mathematics needed to understand the fundamental properties of Laplace Transforms is analysis and, to a lesser extent the normed vector space. Analysis, in particular integration, is needed from the start as it governs the existence conditions for the Laplace Transform itself; however as is soon apparent, calculations involving Laplace Transforms can take place without explicit knowledge of analysis. Normed vector spaces and associated linear algebra put the Laplace Transform on a firm theoretical footing, but can be left until a little later in a book aimed at second year undergraduate mathematics students.

1.2 The Laplace Transform

The definition of the Laplace Transform could hardly be more straightforward. Given a suitable function $F(t)$ the Laplace Transform, written $f(s)$ is defined by

$$f(s) = \int_0^\infty F(t)e^{-st}dt.$$

This bald statement may satisfy most engineers, but not mathematicians. The question of what constitutes a "suitable function" will now be addressed. The integral on the right has infinite range and hence is what is called an improper integral. This too needs careful handling. The notation $\mathcal{L}\{F(t)\}$ is used to denote the Laplace Transform of the function $F(t)$.

Another way of looking at the Laplace Transform is as a mapping from points in the t domain to points in the s domain. Pictorially, Figure 1.1 indicates this mapping process. The time domain t will contain all those functions $F(t)$ whose Laplace Transform exists, whereas the frequency domain s contains all the images $\mathcal{L}\{F(t)\}$. Another aspect of Laplace Transforms that needs mentioning at this stage is that the variable s often has to take complex values. This means that $f(s)$ is a function of a complex variable, which in turn places restrictions on the (real) function $F(t)$ given that the improper integral must converge. Much of the analysis involved in dealing with the image of the function $F(t)$ in the s plane is therefore complex analysis which may be quite new to some readers.

As has been said earlier, engineers are quite happy to use Laplace Transforms to help solve a variety of problems without questioning the convergence of the improper integrals. This goes for some applied mathematicians too. The argument seems to be on the lines that if it gives what looks a reasonable answer, then fine! In our view, this takes the engineer's maxim "if it ain't broke, don't fix it" too far. This is primarily a mathematics textbook, therefore in this opening chapter we shall be more mathematically explicit than is customary in books on Laplace Transforms. In Chapter 4 there is some more pure mathematics when Fourier series are introduced. That is there for similar reasons. One mathematical question that ought to be asked concerns uniqueness. Given a function $F(t)$, its Laplace Transform is surely unique from the well defined nature of the improper integral. However, is it possible for two different functions to have the same Laplace Transform? To put the question a different but equivalent way, is there a function $N(t)$, not identically zero, whose Laplace Transform is zero? For this function, called a *null* function, could be added to any suitable function and the Laplace Transform would remain unchanged. Null functions do exist, but as long as we restrict ourselves to piecewise continuous functions this ceases to be a problem. Here is the definition of piecewise continuous:

Definition 1.1 *If an interval $[0, t_0]$ say can be partitioned into a finite number of subintervals $[0, t_1], [t_1, t_2], [t_2, t_3], \ldots, [t_n, t_0]$ with $0, t_1, t_2, \ldots, t_n, t_0$ an increasing sequence of times and such that a given function $f(t)$ is continuous in each of these subintervals but not necessarily at the end points themselves, then $f(t)$ is piecewise continuous in the interval $[0, t_0]$.*

Only functions that differ at a finite number of points have the same Laplace Transform. If $F_1(t) = F(t)$ except at a finite number of points where they differ by finite values then $\mathcal{L}\{F_1(t)\} = \mathcal{L}\{F(t)\}$. We mention this again in the next chapter when the inverse Laplace Transform is defined.

In this section, we shall examine the conditions for the existence of the Laplace Transform in more detail than is usual. In engineering texts, the simple definition followed by an explanation of exponential order is all that is required. Those that are satisfied with this can virtually skip the next few paragraphs and go on study the elementary properties, Section 1.3. However, some may need to know enough background in terms of the integrals, and so we devote a little space to some fundamentals. We will need to introduce improper integrals, but let us first define the Riemann integral. It is the integral we know and love, and is defined in terms of limits of sums. The strict definition runs as follows:-

Let $F(x)$ be a function which is defined and is bounded in the interval $a \leq x \leq b$ and suppose that m and M are respectively the lower and upper bounds of $F(x)$ in this interval (written $[a, b]$ see Appendix C). Take a set of points

$$x_0 = a, x_1, x_2, \ldots, x_{r-1}, x_r, \ldots, x_n = b$$

and write $\delta_r = x_r - x_{r-1}$. Let M_r, m_r be the bounds of $F(x)$ in the subinterval (x_{r-1}, x_r) and form the sums

$$S = \sum_{r=1}^{n} M_r \delta_r$$

$$s = \sum_{r=1}^{n} m_r \delta_r.$$

These are called respectively the upper and lower Riemann sums corresponding to the mode of subdivision. It is certainly clear that $S \geq s$. There are a variety of ways that can be used to partition the interval (a, b) and each way will have (in general) different M_r and m_r leading to different S and s. Let M be the minimum of all possible M_r and m be the maximum of all possible m_r A lower bound or supremum for the set S is therefore $M(b - a)$ and an upper bound or infimum for the set s is $m(b - a)$. These bounds are of course rough. There are *exact* bounds for S and s, call them J and I respectively. If $I = J$, $F(x)$ is said to be Riemann integrable in (a, b) and the value of the integral is I or J and is denoted by

$$I = J = \int_a^b F(x) dx.$$

For the purist it turns out that the Riemann integral is not quite general enough, and the Stieltjes integral is actually required. However, we will not use this concept which belongs securely in specialist final stage or graduate texts.

The improper integral is defined in the obvious way by taking the limit:

$$\lim_{R \to \infty} \int_a^R F(x) dx = \int_0^\infty F(x) dx$$

provided $F(x)$ is continuous in the interval $a \leq x \leq R$ for every R, and the limit on the left exists.

This is enough of general theory, we now apply it to the Laplace Transform. The parameter x is defined to take the increasing values from 0 to ∞. The condition $|F(x)| \leq Me^{\alpha x}$ is termed "$F(x)$ is of exponential order" and is, speaking loosely, quite a weak condition. All polynomial functions and (of course) exponential functions of the type e^{kx} (k constant) are included as well as bounded functions. Excluded functions are those that have singularities such as $\ln(x)$ or $1/(x - 1)$ and functions that have a growth rate more rapid than exponential, for example e^{x^2}. Functions that have a finite number of finite discontinuities are also included. These have a special role in the theory of Laplace Transforms (see Chapter 3) so we will not dwell on them here: suffice to say that a function such as

$$F(x) = \begin{cases} 1 & 2n < x < 2n + 1 \\ 0 & 2n + 1 < x < 2n + 2 \end{cases} \quad \text{where } n = 0, 1, \ldots$$

is one example. However, the function

$$F(x) = \begin{cases} 1 & x \text{ rational} \\ 0 & x \text{ irrational} \end{cases}$$

is excluded because although all the discontinuities are finite, there are infinitely many of them.

We shall now follow standard practice and use t (time) instead of x as the dummy variable.

1.3   Elementary Properties

The Laplace Transform has many interesting and useful properties, the most fundamental of which is linearity. It is linearity that enables us to add results together to deduce other more complicated ones and is so basic that we state it as a theorem and prove it first.

Theorem 1.2 (Linearity) *If $F_1(t)$ and $F_2(t)$ are two functions whose Laplace Transform exists, then*

$$\mathcal{L}\{aF_1(t) + bF_2(t)\} = a\mathcal{L}\{F_1(t)\} + b\mathcal{L}\{F_2(t)\}$$

where a and b are arbitrary constants.

Proof

$$\begin{aligned}
\mathcal{L}\{aF_1(t) + bF_2(t)\} &= \int_0^\infty (aF_1 + bF_2)e^{-st}dt \\
&= \int_0^\infty \left(aF_1 e^{-st} + bF_2 e^{-st}\right) dt \\
&= a\int_0^\infty F_1 e^{-st}dt + b\int_0^\infty F_2 e^{-st}dt \\
&= a\mathcal{L}\{F_1(t)\} + b\mathcal{L}\{F_2(t)\}
\end{aligned}$$

where we have assumed that

$$|F_1| \leq M_1 e^{\alpha_1 t} \text{ and } |F_2| \leq M_2 e^{\alpha_2 t}$$

so that

$$\begin{aligned}
|aF_1 + bF_2| &\leq |a||F_1| + |b||F_2| \\
&\leq (|a|M_1 + |b|M_2)e^{\alpha_3 t}
\end{aligned}$$

where $\alpha_3 = max\{\alpha_1, \alpha_2\}$. This proves the theorem.

□

In this section, we shall concentrate on those properties of the Laplace Transform that do not involve the calculus. The first of these takes the form of another theorem because of its generality.

Theorem 1.3 (First Shift Theorem) *If it is possible to choose constants M and α such that $|F(t)| \leq M e^{\alpha t}$, that is $F(t)$ is of exponential order, then*

$$\mathcal{L}\{e^{-bt}F(t)\} = f(s+b)$$

provided $b \leq \alpha$. (In practice if $F(t)$ is of exponential order then the constant α can be chosen such that this inequality holds.)

Proof The proof is straightforward and runs as follows:-

$$\begin{aligned}
\mathcal{L}\{e^{-bt}F(t)\} &= \lim_{T \to \infty} \int_0^T e^{-st}e^{-bt}F(t)dt \\
&= \int_0^\infty e^{-st}e^{-bt}F(t)dt \text{ (as the limit exists)} \\
&= \int_0^\infty e^{-(s+b)t}F(t)dt \\
&= f(s+b).
\end{aligned}$$

This establishes the theorem.

$\square$

We shall make considerable use of this once we have established a few elementary Laplace Transforms. This we shall now proceed to do.

Example 1.4 *Find the Laplace Transform of the function $F(t) = t$.*

Solution Using the definition of Laplace Transform,

$$\mathcal{L}(t) = \lim_{T \to \infty} \int_0^T te^{-st}dt.$$

Now, we have that

$$\begin{aligned}
\int_0^T te^{-st}dt &= \left[-\frac{t}{s}e^{-st}\right]_0^T - \int_0^T -\frac{1}{s}e^{-st}dt \\
&= -\frac{T}{s}e^{-sT} + \left[-\frac{1}{s^2}e^{-st}\right]_0^T \\
&= -\frac{T}{s}e^{-sT} - \frac{1}{s^2}e^{-sT} + \frac{1}{s^2}
\end{aligned}$$

this last expression tends to $\dfrac{1}{s^2}$ as $T \to \infty$.

    Hence we have the result

$$\mathcal{L}(t) = \frac{1}{s^2}.$$

We can generalise this result to deduce the following result:
Corollary

$$\mathcal{L}(t^n) = \frac{n!}{s^{n+1}}, \quad n \text{ a positive integer.}$$

Proof The proof is straightforward:

$$\mathcal{L}(t^n) = \int_0^\infty t^n e^{-st} dt \quad \text{this time taking the limit straight away}$$

$$= \left[-\frac{t^n}{s} e^{-st} \right]_0^\infty + \int_0^\infty \frac{n t^{n-1}}{s} e^{-st} dt$$

$$= \frac{n}{s} \mathcal{L}(t^{n-1}).$$

If we put $n = 2$ in this recurrence relation we obtain

$$\mathcal{L}(t^2) = \frac{2}{s} \mathcal{L}(t) = \frac{2}{s^3}.$$

If we assume

$$\mathcal{L}(t^n) = \frac{n!}{s^{n+1}}$$

then

$$\mathcal{L}(t^{n+1}) = \frac{n+1}{s} \frac{n!}{s^{n+1}} = \frac{(n+1)!}{s^{n+2}}.$$

This establishes that

$$\mathcal{L}(t^n) = \frac{n!}{s^{n+1}}$$

by induction.

□

Example 1.5 *Find the Laplace Transform of $\mathcal{L}\{te^{at}\}$ and deduce the value of $\mathcal{L}\{t^n e^{at}\}$, where a is a real constant and n a positive integer.*

Solution Using the first shift theorem with $b = -a$ gives

$$\mathcal{L}\{F(t)e^{at}\} = f(s-a)$$

so with

$$F(t) = t \text{ and } f = \frac{1}{s^2}$$

we get

$$\mathcal{L}\{te^{at}\} = \frac{1}{(s-a)^2}.$$

Using $F(t) = t^n$ the formula

$$\mathcal{L}\{t^n e^{at}\} = \frac{n!}{(s-a)^{n+1}}$$

follows.

Later, we shall generalise this formula further, extending to the case where n is not an integer.

We move on to consider the Laplace Transform of trigonometric functions. Specifically, we shall calculate $\mathcal{L}\{\sin(t)\}$ and $\mathcal{L}\{\cos(t)\}$. It is unfortunate, but the Laplace Transform of the other common trigonometric functions tan, cot, csc and sec do not exist as they all have singularities for finite t. The condition that the function $F(t)$ has to be of exponential order is not obeyed by any of these singular trigonometric functions as can be seen, for example, by noting that

$$|e^{-at}\tan(t)| \to \infty \text{ as } t \to \pi/2$$

and

$$|e^{-at}\cot(t)| \to \infty \text{ as } t \to 0$$

for all values of the constant a. Similarly neither csc nor sec are of exponential order.

In order to find the Laplace Transform of $\sin(t)$ and $\cos(t)$ it is best to determine $\mathcal{L}(e^{it})$ where $i = \sqrt{(-1)}$. The function e^{it} is complex valued, but it is both continuous and bounded for all t so its Laplace Transform certainly exists. Taking the Laplace Transform,

$$
\begin{aligned}
\mathcal{L}(e^{it}) &= \int_0^\infty e^{-st}e^{it}dt \\
&= \int_0^\infty e^{t(i-s)}dt \\
&= \left[\frac{e^{(i-s)t}}{i-s}\right]_0^\infty \\
&= \frac{1}{s-i} \\
&= \frac{s}{s^2+1} + i\frac{1}{s^2+1}.
\end{aligned}
$$

Now,

$$
\begin{aligned}
\mathcal{L}(e^{it}) &= \mathcal{L}(\cos(t) + i\sin(t)) \\
&= \mathcal{L}(\cos(t)) + i\mathcal{L}(\sin(t)).
\end{aligned}
$$

Equating real and imaginary parts gives the two results

$$\mathcal{L}(\cos(t)) = \frac{s}{s^2+1}$$

and

$$\mathcal{L}(\sin(t)) = \frac{1}{s^2+1}.$$

The linearity property has been used here, and will be used in future without further comment.

Given that the restriction on the type of function one can Laplace Transform is weak, i.e. it has to be of exponential order and have at most a finite number of finite jumps, one can find the Laplace Transform of any polynomial, any combination of polynomial with sinusoidal functions and combinations of these with exponentials (provided the exponential functions grow at a rate $\leq e^{at}$ where a is a constant). We can therefore approach the problem of calculating the Laplace Transform of power series. It is possible to take the Laplace Transform of a power series term by term as long as the series uniformly converges to a piecewise continuous function. We shall investigate this further later in the text; meanwhile let us look at the Laplace Transform of functions that are not even continuous.

Functions that are not continuous occur naturally in branches of electrical and control engineering, and in the software industry. One only has to think of switches to realise how widespread discontinuous functions are throughout electronics and computing.

Example 1.6 *Find the Laplace Transform of the function represented by $F(t)$ where*

$$F(t) = \begin{cases} t & 0 \leq t < t_0 \\ 2t_0 - t & t_0 \leq t \leq 2t_0 \\ 0 & t > 2t_0. \end{cases}$$

Solution This function is of the "saw-tooth" variety that is quite common in electrical engineering. There is no question that it is of exponential order and that

$$\int_0^\infty e^{-st} F(t) dt$$

exists and is well defined. $F(t)$ is continuous but not differentiable. This is not troublesome. Carrying out the calculation is a little messy and the details can be checked using computer algebra.

$$
\begin{aligned}
\mathcal{L}(F(t)) &= \int_0^\infty e^{-st} F(t) dt \\
&= \int_0^{t_0} t e^{-st} dt + \int_{t_0}^{2t_0} (2t_0 - t) e^{-st} dt \\
&= \left[-\frac{t}{s} e^{-st} \right]_0^{t_0} + \int_0^{t_0} \frac{1}{s} e^{-st} dt + \left[-\frac{2t_0 - t}{s} e^{-st} \right]_{t_0}^{2t_0} - \int_{t_0}^{2t_0} \frac{1}{s} e^{-st} dt \\
&= -\frac{t_0}{s} e^{-st_0} - \frac{1}{s^2} \left[e^{-st} \right]_0^{t_0} + \frac{t_0}{s} e^{-st_0} + \frac{1}{s^2} \left[e^{-st} \right]_{t_0}^{2t_0} \\
&= \frac{1}{s^2} \left[e^{-st_0} - 1 \right] + \frac{1}{s^2} \left[e^{-2st_0} - e^{-st_0} \right] \\
&= \frac{1}{s^2} \left[1 - 2e^{-st_0} + e^{-2st_0} \right] \\
&= \frac{1}{s^2} \left[1 - e^{-st_0} \right]^2
\end{aligned}
$$

$$= \frac{4}{s^2} e^{-st_0} \sinh^2(\frac{1}{2} st_0).$$

In the next chapter we shall investigate in more detail the properties of discontinuous functions such as the Heaviside unit step function. As an introduction to this, let us do the following example.

Example 1.7 *Determine the Laplace Transform of the step function $F(t)$ defined by*

$$F(t) = \begin{cases} 0 & 0 \le t < t_0 \\ a & t \ge t_0. \end{cases}$$

Solution $F(t)$ itself is bounded, so there is no question that it is also of exponential order. The Laplace Transform of $F(t)$ is therefore

$$\mathcal{L}(F(t)) = \int_0^\infty e^{-st} F(t) dt$$

$$= \int_{t_0}^\infty ae^{-st} dt$$

$$= \left[-\frac{a}{s} e^{-st} \right]_{t_0}^\infty$$

$$= \frac{a}{s} e^{-st_0}.$$

Here is another useful result.

Theorem 1.8 *If $\mathcal{L}(F(t)) = f(s)$ then $\mathcal{L}(tF(t)) = -\dfrac{d}{ds} f(s)$ and in general $\mathcal{L}(t^n F(t)) = (-1)^n \dfrac{d^n}{ds^n} f(s)$.*

Proof Let us start with the definition of Laplace Transform

$$\mathcal{L}(F(t)) = \int_0^\infty e^{-st} F(t) dt$$

and differentiate this with respect to s to give

$$\frac{df}{ds} = \frac{d}{ds} \int_0^\infty e^{-st} F(t) dt$$

$$= \int_0^\infty -te^{-st} F(t) dt$$

assuming absolute convergence to justify interchanging differentiation and (improper) integration. Hence

$$\mathcal{L}(tF(t)) = -\frac{d}{ds} f(s).$$

One can now see how to progress by induction. Assume the result holds for n, so that

$$\mathcal{L}(t^n F(t)) = (-1)^n \frac{d^n}{ds^n} f(s)$$

and differentiate both sides with respect to s (assuming all appropriate convergence properties) to give

$$\int_0^\infty -t^{n+1} e^{-st} F(t) dt = (-1)^n \frac{d^{n+1}}{ds^{n+1}} f(s)$$

or

$$\int_0^\infty t^{n+1} e^{-st} F(t) dt = (-1)^{n+1} \frac{d^{n+1}}{ds^{n+1}} f(s).$$

So

$$\mathcal{L}(t^{n+1} F(t)) = (-1)^{n+1} \frac{d^{n+1}}{ds^{n+1}} f(s)$$

which establishes the result by induction.

□

Example 1.9 *Determine the Laplace Transform of the function* $t \sin(t)$.

Solution To evaluate this Laplace Transform we use Theorem 1.8 with $f(t) = \sin(t)$. This gives

$$\mathcal{L}\{t \sin(t)\} = -\frac{d}{ds}\left\{\frac{1}{1+s^2}\right\} = \frac{2s}{(1+s^2)^2}$$

which is the required result.

1.4   Exercises

1. For each of the following functions, determine which has a Laplace Transform. If it exists, find it; if it does not, say briefly why.

   (a) $\ln(t)$, (b) e^{3t}, (c) e^{t^2}, (d) $e^{1/t}$, (e) $1/t$,

   (f) $f(t) = \begin{cases} 1 & \text{if } t \text{ is even} \\ 0 & \text{if } t \text{ is odd.} \end{cases}$

2. Determine from first principles the Laplace Transform of the following functions:-

   (a) e^{kt}, (b) t^2, (c) $\cosh(t)$.

3. Find the Laplace Transforms of the following functions:-

(a) $t^2 e^{-3t}$, (b) $4t + 6e^{4t}$, (c) $e^{-4t} \sin(5t)$.

4. Find the Laplace Transform of the function $F(t)$, where $F(t)$ is given by
$$F(t) = \begin{cases} t & 0 \leq t < 1 \\ 2 - t & 1 \leq t < 2 \\ 0 & \text{otherwise.} \end{cases}$$

5. Use the property of Theorem 1.8 to determine the following Laplace Transforms

(a) te^{2t}, (b) $t\cos(t)$, (c) $t^2 \cos(t)$.

6. Find the Laplace Transforms of the following functions:-

(a) $\sin(\omega t + \phi)$, (b) $e^{5t} \cosh(6t)$.

7. If $G(at + b) = F(t)$ determine the Laplace Transform of G in terms of $\mathcal{L}\{F\} = \bar{f}(s)$ and a finite integral.

8. Prove the following change of scale result:-
$$\mathcal{L}\{F(at)\} = \frac{1}{a} f\left(\frac{s}{a}\right).$$
Hence evaluate the Laplace Transforms of the two functions

(a) $t\cos(6t)$, (b) $t^2 \cos(7t)$.

2

Further Properties of the Laplace Transform

2.1   Real Functions

Sometimes, a function $F(t)$ represents a natural or engineering process that has no obvious starting value. Statisticians call this a *time series*. Although we shall not be considering $F(t)$ as stochastic, it is nevertheless worth introducing a way of "switching on" a function. Let us start by finding the Laplace Transform of a step function the name of which pays homage to the pioneering electrical engineer Oliver Heaviside (1850–1925). The formal definition runs as follows.

Definition 2.1 *Heaviside's Unit Step Function, or simply the unit step function, is defined as*

$$H(t) = \left\{ \begin{array}{ll} 0 & t < 0 \\ 1 & t \geq 0. \end{array} \right.$$

Since $H(t)$ is precisely the same as 1 for $t > 0$, the Laplace Transform of $H(t)$ must be the same as the Laplace Transform of 1, i.e. $1/s$. The switching on of an arbitrary function is achieved simply by multiplying it by the standard function $H(t)$, so if $F(t)$ is given by the function shown in Figure 2.1 and we multiply this function by the Heaviside Unit Step function $H(t)$ to obtain $H(t)F(t)$, Figure 2.2 results. Examples of the use of $H(t)$ are to be found throughout the rest of this text, in particular Sections 2.3 and 3.4. Sometimes it is necessary to define what is called the *two sided* Laplace Transform

$$\int_{-\infty}^{\infty} e^{-st} F(t) dt$$

13

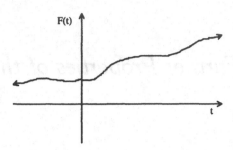

Figure 2.1: $F(t)$, a function with no well defined starting value

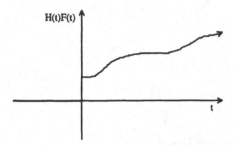

Figure 2.2: $H(t)F(t)$, the function is now zero before $t = 0$

which makes a great deal of mathematical sense. However the additional problems that arise by allowing negative values of t are severe and limit the use of the two sided Laplace Transform. For this reason, the two sided transform will not be pursued here.

2.2 Derivative Property of the Laplace Transform

Suppose a differentiable function $F(t)$ has Laplace Transform $f(s)$, we can find the Laplace Transform

$$\mathcal{L}\{F'(t)\} = \int_0^\infty e^{-st} F'(t) dt$$

of its derivative $F'(t)$ through the following theorem.

Theorem 2.2

$$\mathcal{L}\{F'(t)\} = \int_0^\infty e^{-st} F'(t) dt = -F(0) + sf(s).$$

Proof Integrating by parts once gives

$$\mathcal{L}\{F'(t)\} = \left[F(t)e^{-st}\right]_0^\infty + \int_0^\infty se^{-st}F(t)dt$$

$$= -F(0) + sf(s)$$

where $F(0)$ is the value of $F(t)$ at $t = 0$.

□

This is an important result and lies behind future applications that involve solving linear differential equations. The key property is that the transform of a derivative $F'(t)$ does not itself involve a derivative, only $-F(0) + sf(s)$ which is an algebraic expression involving $f(s)$. The downside is that the value $F(0)$ is required. Effectively, an integration has taken place and the constant of integration is $F(0)$. In the next chapter, this is exploited further through solving differential equations. Later still in this text, partial differential equations are solved. Let us proceed here by finding the Laplace Transform of the second derivative of $F(t)$. We also state this in the form of a theorem.

Theorem 2.3 *If $F(t)$ is a twice differentiable function of t then*

$$\mathcal{L}\{F''(t)\} = s^2 f(s) - sF(0) - F'(0).$$

Proof The proof is unremarkable and involves integrating by parts twice. Here are the details.

$$\mathcal{L}\{F''(t)\} = \int_0^\infty e^{-st}F''(t)dt$$

$$= \left[F'(t)e^{-st}\right]_0^\infty + \int_0^\infty se^{-st}F'(t)dt$$

$$= -F'(0) + \left[sF(t)e^{-st}\right]_0^\infty + \int_0^\infty s^2e^{-st}F(t)dt$$

$$= -F'(0) - sF(0) + s^2 f(s)$$

$$= s^2 f(s) - sF(0) - F'(0).$$

□

The general result, proved by induction, is

$$\mathcal{L}\{F^{(n)}(t)\} = s^n f(s) - s^{n-1}F(0) - s^{n-2}F'(0) - \cdots - F^{(n-1)}(0)$$

where n is a positive integer. Note the appearance of n constants on the right hand side. This of course is the result of integrating this number of times.

This result, as we have said, has wide application so it is worth getting to know. Consider the result

$$\mathcal{L}(\sin(\omega t)) = \frac{\omega}{s^2 + \omega^2}.$$

Now,

$$\frac{d}{dt}(\sin(\omega t)) = \omega \cos(\omega t)$$

so using the formula

$$\mathcal{L}(F'(t)) = sf(s) - F(0)$$

with $F(t) = \sin(\omega t)$ we have

$$\mathcal{L}\{\omega \cos(\omega t)\} = s\frac{\omega}{s^2 + \omega^2} - 0$$

so

$$\mathcal{L}\{\cos(\omega t)\} = \frac{s}{s^2 + \omega^2}$$

another standard result.

Another appropriate question to tackle at this point is the determination of the value of the Laplace Transform of

$$\int_0^t F(u)du.$$

First of all, the function $F(t)$ must be integrable in such a way that

$$g(t) = \int_0^t F(u)du$$

is of exponential order. From this definition of $g(t)$ it is immediately apparent that $g(0) = 0$ and that $g'(t) = F(t)$. This latter result is called the fundamental theorem of the calculus. We can now use the result

$$\mathcal{L}\{g'(t)\} = s\overline{g}(s) - g(0)$$

to obtain

$$\mathcal{L}\{F(t)\} = f(s) = s\overline{g}(s)$$

where we have written $\mathcal{L}\{g(t)\} = \overline{g}(s)$. Hence

$$\overline{g}(s) = \frac{f(s)}{s}$$

which finally gives the result

$$\mathcal{L}\left(\int_0^t F(u)du\right) = \frac{f(s)}{s}.$$

The following result is also useful and can be stated in the form of a theorem.

Theorem 2.4 *If $\mathcal{L}(F(t)) = f(s)$ then $\mathcal{L}\left\{\dfrac{F(t)}{t}\right\} = \displaystyle\int_s^\infty f(u)du$, assuming that*

$$\mathcal{L}\left\{\frac{F(t)}{t}\right\} \to 0 \text{ as } s \to \infty.$$

Proof Let $G(t)$ be the function $F(t)/t$, so that $F(t) = tG(t)$. Using the property

$$\mathcal{L}\{tG(t)\} = -\frac{d}{ds}\mathcal{L}\{G(t)\}$$

we deduce that

$$f(s) = \mathcal{L}\{F(t)\} = -\frac{d}{ds}\mathcal{L}\left\{\frac{F(t)}{t}\right\}.$$

Integrating both sides of this with respect to s from s to ∞ gives

$$\int_s^\infty f(u)du = \left[-\mathcal{L}\left\{\frac{F(t)}{t}\right\}\right]_s^\infty = \mathcal{L}\left\{\frac{F(t)}{t}\right\}|_s = \mathcal{L}\left\{\frac{F(t)}{t}\right\}$$

since

$$\mathcal{L}\left\{\frac{F(t)}{t}\right\} \to 0 \text{ as } s \to \infty$$

which completes the proof.

□

The function

$$Si(t) = \int_0^t \frac{\sin(u)}{u}du$$

defines the Sine Integral function which occurs in the study of optics. The formula for its Laplace Transform can now be easily derived as follows.

Let $F(t) = \sin(t)$ in the result

$$\mathcal{L}\left(\frac{F(t)}{t}\right) = \int_s^\infty f(u)du$$

to give

$$\mathcal{L}\left(\frac{\sin(t)}{t}\right) = \int_s^\infty \frac{du}{1+u^2}$$

$$= \left[\tan^{-1}(u)\right]_s^\infty$$

$$= \frac{\pi}{2} - \tan^{-1}(s)$$

$$= \tan^{-1}\left(\frac{1}{s}\right).$$

We now use the result

$$\mathcal{L}\left(\int_0^t F(u)du\right) = \frac{f(s)}{s}$$

to deduce that

$$\mathcal{L}\left(\int_0^t \frac{\sin(u)}{u}du\right) = \mathcal{L}\{Si(t)\} = \frac{1}{s}\tan^{-1}\left(\frac{1}{s}\right).$$

2.3   Heaviside's Unit Step Function

As promised earlier, we devote this section to exploring some properties of Heaviside's Unit Step Function $H(t)$. The Laplace Transform of $H(t)$ has already been shown to be the same as the Laplace Transform of 1, i.e. $1/s$. The Laplace Transform of $H(t - t_0)$, $t_0 > 0$, is a little more enlightening:

$$\mathcal{L}\{H(t - t_0)\} = \int_0^\infty H(t - t_0)e^{-st}dt.$$

Now, since $H(t - t_0) = 0$ for $t < t_0$ this Laplace Transform is

$$\mathcal{L}\{H(t - t_0)\} = \int_{t_0}^\infty e^{-st}dt = \left[-\frac{e^{-st}}{s}\right]_{t_0}^\infty = \frac{e^{-st_0}}{s}.$$

This result is generalised through the following theorem.

Theorem 2.5 (Second Shift Theorem) *If $F(t)$ is a function of exponential order in t then*

$$\mathcal{L}\{H(t - t_0)F(t - t_0)\} = e^{-st_0}f(s)$$

where $f(s)$ is the Laplace Transform of $F(t)$.

Proof This result is proved by direct integration.

$$\begin{aligned}
\mathcal{L}\{H(t - t_0)F(t - t_0)\} &= \int_0^\infty H(t - t_0)F(t - t_0)e^{-st}dt \\
&= \int_{t_0}^\infty F(t - t_0)e^{-st}dt \text{ (by definition of } H) \\
&= e^{-st_0}\int_0^\infty F(u)e^{-s(u+t_0)}du \text{ (writing } u = t - t_0) \\
&= e^{-st_0}f(s).
\end{aligned}$$

This establishes the theorem.

□

The only condition on $F(t)$ is that it is a function that is of exponential order which means of course that it is free from singularities for $t > t_0$. The principal use of this theorem is that it enables us to determine the Laplace Transform of a function that is switched on at time $t = t_0$. Here is a straightforward example.

Example 2.6 *Determine the Laplace Transform of the sine function switched on at time $t = 3$.*

Solution The sine function required that starts at $t = 3$ is $S(t)$ where

$$S(t) = \begin{cases} \sin(t) & t \geq 3 \\ 0 & t < 3. \end{cases}$$

We can use the Heaviside Step function to write

$$S(t) = H(t-3)\sin(t).$$

The Second Shift Theorem can then be used by utilising the summation formula

$$\sin(t) = \sin(t-3+3) = \sin(t-3)\cos(3) + \cos(t-3)\sin(3)$$

so

$$\mathcal{L}\{S(t)\} = \mathcal{L}\{H(t-3)\sin(t-3)\}\cos(3) + \mathcal{L}\{H(t-3)\cos(t-3)\}\sin(3).$$

This may seem a strange step to take, but in order to use the Second Shift Theorem it is essential to get the arguments of both the Heaviside function and the target function in the question the same. We can now use the Second Shift Theorem directly to give

$$\mathcal{L}\{S(t)\} = e^{-3s}\cos(3)\frac{1}{s^2+1} + e^{-3s}\sin(3)\frac{s}{s^2+1}$$

or

$$\mathcal{L}\{S(t)\} = (\cos 3 + s\sin 3)e^{-3s}/(s^2+1).$$

2.4 Inverse Laplace Transform

Virtually all operations have inverses. Addition has subtraction, multiplication has division, differentiation has integration. The Laplace Transform is no exception, and we can define the *Inverse* Laplace Transform as follows.

Definition 2.7 *If $F(t)$ has the Laplace Transform $f(s)$, that is*

$$\mathcal{L}\{F(t)\} = f(s)$$

then the Inverse Laplace Transform is defined by

$$\mathcal{L}^{-1}\{f(s)\} = F(t)$$

and is unique apart from null functions.

Perhaps the most important property of the inverse transform to establish is its linearity. We state this as a theorem.

Theorem 2.8 *The Inverse Laplace Transform is linear, i.e.*

$$\mathcal{L}^{-1}\{af_1(s) + bf_2(s)\} = a\mathcal{L}^{-1}\{f_1(s)\} + b\mathcal{L}^{-1}\{f_2(s)\}.$$

Proof Linearity is easily established as follows. Since the Laplace Transform is linear, we have for suitably well behaved functions $F_1(t)$ and $F_2(t)$:

$$\mathcal{L}\{aF_1(t) + bF_2(t)\} = a\mathcal{L}\{F_1(t)\} + b\mathcal{L}\{F_2(t)\} = af_1(s) + bf_2(s).$$

Taking the Inverse Laplace Transform of this expression gives

$$aF_1(t) + bF_2(t) = \mathcal{L}^{-1}\{af_1(s) + bf_2(s)\}$$

which is the same as

$$a\mathcal{L}^{-1}\{f_1(s)\} + b\mathcal{L}^{-1}\{f_2(s)\} = \mathcal{L}^{-1}\{af_1(s) + bf_2(s)\}$$

and this has established linearity of $\mathcal{L}^{-1}\{f(s)\}$.

□

Another important property is uniqueness. In Chapter 1 we mentioned that the Laplace Transform was indeed unique apart from null functions (functions whose Laplace Transform is zero). It follows immediately that the Inverse Laplace Transform is also unique apart from the possible addition of null functions. These take the form of isolated values and can be discounted for all practical purposes.

As is quite common with inverse operations there is no systematic method of determining Inverse Laplace Transforms. The calculus provides a good example where there are plenty of systematic rules for differentiation: the product rule, the quotient rule, the chain rule. However by contrast there are no systematic rules for the inverse operation, integration. If we have an integral to find, we may try substitution or integration by parts, but there is no guarantee of success. Indeed, the integral may not be possible in terms of elementary functions. Derivatives that exist can always be found by using the rules; this is not so for integrals. The situation regarding the Laplace Transform is not quite the same in that it may not be possible to find $\mathcal{L}\{F(t)\}$ explicitly because it is an integral. There is certainly no guarantee of being able to find $\mathcal{L}^{-1}\{f(s)\}$ and we have to devise various methods of trying so to do. For example, given an arbitrary function of s there is no guarantee whatsoever that a function of t can be found that is its Inverse Laplace Transform. One necessary condition for example is that the function of s must tend to zero as $s \to \infty$. When we are certain that a function of s has arisen from a Laplace Transform, there are techniques and theorems that can help us invert it. Partial fractions simplify rational functions and can help identify standard forms (the exponential and trigonometric functions for example), then there are the shift theorems which we have just met which extend further the repertoire of standard forms. Engineering texts spend a considerable amount of space building up a library of specific Inverse Laplace Transforms and to ways of extending these via the calculus. To a certain extent we need to do this too. Therefore we next do some reasonably elementary examples.

Example 2.9 *Use partial fractions to determine*

$$\mathcal{L}^{-1}\left\{\frac{a}{s^2 - a^2}\right\}.$$

Solution Noting that

$$\frac{a}{s^2 - a^2} = \frac{1}{2}\left[\frac{1}{s-a} - \frac{1}{s+a}\right]$$

gives straight away that

$$\mathcal{L}^{-1}\left\{\frac{a}{s^2 - a^2}\right\} = \frac{1}{2}(e^{at} - e^{-at}) = \sinh(at).$$

The first shift theorem has been used on each of the functions $1/(s-a)$ and $1/(s+a)$ together with the standard result $\mathcal{L}^{-1}\{1/s\} = 1$. Here is another example.

Example 2.10 *Determine the value of*

$$\mathcal{L}^{-1}\left\{\frac{s^2}{(s+3)^3}\right\}.$$

Solution Noting the standard partial fraction decomposition

$$\frac{s^2}{(s+3)^3} = \frac{1}{s+3} - \frac{6}{(s+3)^2} + \frac{9}{(s+3)^3}$$

we use the first shift theorem on each of the three terms in turn to give

$$\mathcal{L}^{-1}\left\{\frac{s^2}{(s+3)^3}\right\} = \mathcal{L}^{-1}\frac{1}{s+3} - \mathcal{L}^{-1}\frac{6}{(s+3)^2} + \mathcal{L}^{-1}\frac{9}{(s+3)^3}$$

$$= e^{-3t} - 6te^{-3t} + \frac{9}{2}t^2e^{-3t}$$

where we have used the linearity property of the $\mathcal{L}^{-1}$ operator. Finally, we do the following four-in-one example to hone our skills.

Example 2.11 *Determine the following inverse Laplace Transforms*

$(a)\mathcal{L}^{-1}\dfrac{(s+3)}{s(s-1)(s+2)}; (b)\mathcal{L}^{-1}\dfrac{(s-1)}{s^2+2s-8}; (c)\mathcal{L}^{-1}\dfrac{3s+7}{s^2-2s+5}; (d)\mathcal{L}^{-1}\dfrac{e^{-7s}}{(s+3)^3}.$

Solution All of these problems are tackled in a similar way, by decomposing the expression into partial fractions, using shift theorems, then identifying the simplified expressions with various standard forms.

(a) Using partial fraction decomposition and not dwelling on the detail we get

$$\frac{s+3}{s(s-1)(s+2)} = -\frac{3}{2s} + \frac{4}{3(s-1)} + \frac{1}{6(s+2)}.$$

Hence, operating on both sides with the Inverse Laplace Transform operator gives

$$\mathcal{L}^{-1}\frac{s+3}{s(s-1)(s+2)} = -\mathcal{L}^{-1}\frac{3}{2s} + \mathcal{L}^{-1}\frac{4}{3(s-1)} + \mathcal{L}^{-1}\frac{1}{6(s+2)}$$

$$= -\frac{3}{2}\mathcal{L}^{-1}\frac{1}{s} + \frac{4}{3}\mathcal{L}^{-1}\frac{1}{s-1} + \frac{1}{6}\mathcal{L}^{-1}\frac{1}{s+2}$$

using the linearity property of $\mathcal{L}^{-1}$ once more. Finally, using the standard forms, we get

$$\mathcal{L}^{-1}\left\{\frac{s+3}{s(s-1)(s+2)}\right\} = -\frac{3}{2} + \frac{4}{3}e^t + \frac{1}{6}e^{-2t}.$$

(b) The expression

$$\frac{s-1}{s^2+2s-8}$$

is factorised to

$$\frac{s-1}{(s+4)(s-2)}$$

which, using partial fractions is

$$\frac{1}{6(s-2)} + \frac{5}{6(s+4)}.$$

Therefore, taking Inverse Laplace Transforms gives

$$\mathcal{L}^{-1}\frac{s-1}{s^2+2s-8} = \frac{1}{6}e^{2t} + \frac{5}{6}e^{-4t}.$$

(c) The denominator of the rational function

$$\frac{3s+7}{s^2-2s+5}$$

does not factorise. In this case we use completing the square and standard trigonometric forms as follows:

$$\frac{3s+7}{s^2-2s+5} = \frac{3s+7}{(s-1)^2+4} = \frac{3(s-1)+10}{(s-1)^2+4}.$$

So

$$\mathcal{L}^{-1}\frac{3s+7}{s^2-2s+5} = 3\mathcal{L}^{-1}\frac{(s-1)}{(s-1)^2+4} + 5\mathcal{L}^{-1}\frac{2}{(s-1)^2+4}$$

$$= 3e^t\cos(2t) + 5e^t\sin(2t).$$

Again, the first shift theorem has been used.

(d) The final Inverse Laplace Transform is slightly different. The expression

$$\frac{e^{-7s}}{(s-3)^3}$$

contains an exponential in the numerator, therefore it is expected that the second shift theorem will have to be used. There is a little "fiddling" that needs to take place here. First of all, note that

$$\mathcal{L}^{-1}\frac{1}{(s-3)^3} = \frac{1}{2}t^2e^{3t}$$

using the first shift theorem. So

$$\mathcal{L}^{-1}\frac{e^{-7s}}{(s-3)^3} = \left\{ \begin{array}{ll} \frac{1}{2}(t-7)^2 e^{3(t-7)} & t > 7 \\ 0 & 0 \le t \le 7. \end{array} \right.$$

Of course, this can succinctly be expressed using the Heaviside Unit Step Function as

$$\frac{1}{2}H(t-7)(t-7)^2 e^{3(t-7)}.$$

We shall get more practice at this kind of inversion exercise, but you should try your hand at a few of the exercises at the end of the chapter.

2.5   Limiting Theorems

In many branches of mathematics there is a necessity to solve differential equations. Later chapters give details of how some of these equations can be solved by using Laplace Transform techniques. Unfortunately, it is sometimes the case that it is not possible to invert $f(s)$ to retrieve the desired solution to the original problem. Numerical inversion techniques are possible and these can be found in some software packages, especially those used by control engineers. Insight into the behaviour of the solution can be deduced without actually solving the differential equation by examining the asymptotic character of $f(s)$ for small s or large s. In fact, it is often very useful to determine this asymptotic behaviour without solving the equation, even when exact solutions are available as these solutions are often complex and difficult to obtain let alone interpret. In this section two theorems that help us to find this asymptotic behaviour are investigated.

Theorem 2.12 (Initial Value) *If the indicated limits exist then*

$$\lim_{t \to 0} F(t) = \lim_{s \to \infty} sf(s)$$

(The left hand side is $F(0)$ of course, or $F(0+)$ if $\lim_{t \to 0} F(t)$ is not unique.)

Proof We have already established that

$$\mathcal{L}\{F'(t)\} = sf(s) - F(0). \tag{2.1}$$

However, if $F'(t)$ obeys the usual criteria for the existence of the Laplace Transform, that is $F'(t)$ is of exponential order and is piecewise continuous, then

$$\left| \int_0^\infty e^{-st}F'(t)dt \right| \le \int_0^\infty |e^{-st}F'(t)|dt$$

$$\le \int_0^\infty e^{-st}e^{Mt}dt$$

$$= -\frac{1}{M-s} \to 0 \text{ as } s \to \infty.$$

Thus letting $s \to \infty$ in (2.1) yields the result. □

Theorem 2.13 (Final Value) *If the limits indicated exist, then*

$$\lim_{t \to \infty} F(t) = \lim_{s \to 0} sf(s).$$

Proof Again we start with the formula for the Laplace Transform of the derivative of $F(t)$

$$\mathcal{L}\{F'(t)\} = \int_0^\infty e^{-st}F'(t)dt = sf(s) - F(0) \qquad (2.2)$$

this time writing the integral out explicitly. The limit of the integral as $s \to 0$ is

$$\begin{aligned}
\lim_{s \to 0} \int_0^\infty e^{-st}F'(t)dt &= \lim_{s \to 0} \lim_{T \to \infty} \int_0^T e^{-st}F'(t)dt \\
&= \lim_{s \to 0} \lim_{T \to \infty} \{e^{-sT}F(T) - F(0)\} \\
&= \lim_{T \to \infty} F(T) - F(0) \\
&= \lim_{t \to \infty} F(t) - F(0).
\end{aligned}$$

Thus we have, using Equation 2.2,

$$\lim_{t \to \infty} F(t) - F(0) = \lim_{s \to 0} sf(s) - F(0)$$

from which, on cancellation of $-F(0)$, the theorem follows.

$\square$

Since the improper integral converges independently of the value of s and all limits exist (a priori assumption), it is therefore correct to have assumed that the order of the two processes (taking the limit and performing the integral) can be exchanged. (This has in fact been demonstrated explicitly in this proof.)

Suppose that the function $F(t)$ can be expressed as a power series as follows

$$F(t) = a_0 + a_1 t + a_2 t^2 + \cdots + a_n t^n + \cdots.$$

If we assume that the Laplace Transform of $F(t)$ exists, $F(t)$ is of exponential order and is piecewise continuous. If, further, we assume that the power series for $F(t)$ is absolutely and uniformly convergent the Laplace Transform can be applied term by term

$$\begin{aligned}
\mathcal{L}\{F(t)\} = f(s) &= \mathcal{L}\{a_0 + a_1 t + a_2 t^2 + \cdots + a_n t^n + \cdots\} \\
&= a_0\mathcal{L}\{1\} + a_1\mathcal{L}\{t\} + a_2\mathcal{L}\{t^2\} \cdots + a_n\mathcal{L}\{t^n\} + \cdots
\end{aligned}$$

provided the transformed series is convergent. Using the standard form

$$\mathcal{L}\{t^n\} = \frac{n!}{s^{n+1}}$$

the right hand side becomes

$$\frac{a_0}{s} + \frac{a_1}{s^2} + \frac{2a_2}{s^3} + \cdots + \frac{n!a_n}{s^{n+1}} + \cdots.$$

Hence

$$f(s) = \frac{a_0}{s} + \frac{a_1}{s^2} + \frac{2a_2}{s^3} + \cdots + \frac{n!a_n}{s^{n+1}} + \cdots.$$

Example 2.14 *Demonstrate the initial and final value theorems using the function $F(t) = e^{-t}$. Expand e^{-t} as a power series, evaluate term by term and confirm the legitimacy of term by term evaluation.*

Solution

$$\mathcal{L}\{e^{-t}\} = \frac{1}{s+1}$$

$$\lim_{t \to 0} F(t) = F(0) = e^{-0} = 1$$

$$\lim_{s \to \infty} sf(s) = \lim_{s \to \infty} \frac{s}{s+1} = 1.$$

This confirms the initial value theorem. The final value theorem is also confirmed as follows:-

$$\lim_{t \to \infty} F(t) = \lim_{t \to \infty} e^{-t} = 0$$

$$\lim_{s \to 0} sf(s) = \lim_{s \to 0} \frac{s}{s+1} = 0.$$

The power series expansion for e^{-t} is

$$e^{-t} = 1 - t + \frac{t^2}{2!} - \frac{t^3}{3!} + \cdots + (-1)^n \frac{t^n}{n!} + \cdots$$

$$\mathcal{L}\{e^{-t}\} = \frac{1}{s} - \frac{1}{s^2} + \frac{1}{s^3} - \cdots + \frac{(-1)^n}{s^{n+1}} + \cdots$$

$$= \frac{1}{s}\left(1 + \frac{1}{s}\right)^{-1} = \frac{1}{s+1}.$$

Hence the term by term evaluation of the power series expansion for e^{-t} gives the right answer. This is not a proof of the series expansion method of course, merely a verification that the method gives the right answer in this instance.

2.6   The Impulse Function

There is a whole class of "functions" that, strictly, are not functions at all. In order to be a function, an expression has to be defined for all values of the variable in the specified range. When this is not so, then the expression is not a function because it is not well defined. It may not seem at all sensible for us to bother with such creatures, in that if a function is not defined at a certain point then what use is it? However, if a "function" instead of being well defined possesses some global property, then it indeed does turn out to be worth considering such pathological objects. Of course, having taken the decision to consider such objects, strictly there needs to be a whole new mathematical language constructed to deal with them. Notions such as adding them together, multiplying them, performing operations such as integration cannot be done without preliminary mathematics. The general consideration of this kind of object forms the study of *generalised functions* (see Jones (1966) or Lighthill (1970)) which is outside the scope of this text. For our purposes we introduce the first such function which occurred naturally in the field of electrical engineering and is the so called impulse function. It is sometimes called Dirac's δ function after the pioneering theoretical physicist P.A.M. Dirac (1902–1984). It has the following definition which involves its integral. This has not been defined properly, but if we write the definition first we can then comment on the integral.

Definition 2.15 *The Dirac-δ function $\delta(t)$ is defined as having the following properties*

$$\delta(t) = 0 \, \forall t \, , t \neq 0 \tag{2.3}$$

$$\int_{-\infty}^{\infty} h(t)\delta(t)dt = h(0) \tag{2.4}$$

for any function $h(t)$ continuous in $(-\infty, \infty)$.

We shall see in the next paragraph that the Dirac-δ function can be thought of as the limiting case of a top hat function of unit area as it becomes infinitesimally thin but infinitely tall, i.e. the following limit

$$\delta(t) = \lim_{T \to \infty} T_p(t)$$

where

$$T_p(t) = \begin{cases} 0 & t \leq -1/T \\ \frac{1}{2}T & -1/T < t < 1/T \\ 0 & t \geq 1/T. \end{cases}$$

The integral in the definition can then be written as follows:

$$\int_{-\infty}^{\infty} h(t) \lim_{T \to \infty} T_p(t)dt = \lim_{T \to \infty} \int_{-\infty}^{\infty} h(t)T_p(t)dt$$

provided the limits can be exchanged which of course depends on the behaviour of the function $h(t)$ but this can be so chosen to fulfil our needs. The integral inside the limit exists, being the product of continuous functions, and its value is the area under the curve $h(t)T_p(t)$. This area will approach the value $h(0)$ as $T \to \infty$ by the following argument. For sufficiently large values of T, the interval $[-1/T, 1/T]$ will be small enough for the value of $h(t)$ not to differ very much from its value at the origin. In this case we can write $h(t) = h(0) + \epsilon(t)$ where $|\epsilon(t)|$ is in some sense small and tends to zero as $T \to \infty$. The integral thus can be seen to tend to $h(0)$ as $T \to \infty$ and the property is established.

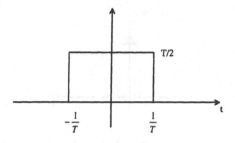

Figure 2.3: The "top hat" function

Returning to the definition of $\delta(t)$ strictly, the first condition is redundant; only the second is necessary, but it is very convenient to retain it. Now as we have said, $\delta(t)$ is not a true function because it has not been defined for $t = 0$. $\delta(0)$ has no value. Equivalent conditions to Equation 2.4 are:-

$$\int_{0-}^{\infty} h(t)\delta(t)dt = h(0)$$

and

$$\int_{-\infty}^{0+} h(t)\delta(t)dt = h(0).$$

These follow from a similar argument as before using a limiting definition of $\delta(t)$ in terms of the top hat function. In this section, wherever the integral of a δ function (or later related "derivatives") occurs it will be assumed to involve this kind of limiting process. The details of taking the limit will however be omitted.

Let us now look at a more visual approach. As we have seen algebraically in the last paragraph $\delta(t)$ is sometimes called the impulse function because it can be thought of as the shape of Figure 2.3, the top hat function if we let $T \to \infty$. Of course there are many shapes that will behave like $\delta(t)$ in some limit. The top hat function is one of the simplest to state and visualise. The crucial property is that the area under this top hat function is unity for all values of T, so letting $T \to \infty$ preserves this property. Diagrammatically, the Dirac-δ or impulse function is represented by an arrow as in Figure 2.4 where the length of the arrow is unity. Using Equation 2.4 with $h \equiv 1$ we see that

$$\int_{-\infty}^{\infty} \delta(t)dt = 1$$

which is consistent with the area under $\delta(t)$ being unity.

We now ask ourselves what is the Laplace Transform of $\delta(t)$? Does it exist? We suspect that it might be 1 for Equation 2.4 with $h(t) = e^{-st}$, a perfectly valid choice of $h(t)$ gives

$$\int_{-\infty}^{\infty} \delta(t)e^{-st}dt = \int_{0-}^{\infty} \delta(t)e^{-st}dt = 1.$$

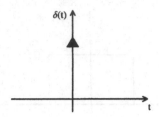

Figure 2.4: The Dirac-δ function

However, we progress with care. This is good advice when dealing with generalised functions. Let us take the Laplace Transform of the top hat function $T_p(t)$ defined mathematically by

$$T_p(t) = \begin{cases} 0 & t \le -1/T \\ \frac{1}{2}T & -1/T < t < 1/T \\ 0 & t \ge 1/T. \end{cases}$$

The calculation proceeds as follows:-

$$\begin{aligned} \mathcal{L}\{T_p(t)\} &= \int_0^\infty T_p(t)e^{-st}dt \\ &= \int_0^{1/T} \frac{1}{2}Te^{-st}dt \\ &= \left[-\frac{T}{2s}e^{-st}\right]_0^{1/T} \\ &= \left[\frac{T}{2s} - \frac{T}{2s}e^{-s/T}\right]. \end{aligned}$$

As $T \to \infty$,

$$e^{-s/T} \approx 1 - \frac{s}{T} + O\left(\frac{1}{T^2}\right)$$

hence

$$\frac{T}{2s} - \frac{T}{2s}e^{-s/T} \approx \frac{1}{2} + O\left(\frac{1}{T}\right)$$

which $\to \frac{1}{2}$ as $T \to \infty$.

In Laplace Transform theory it is usual to define the impulse function $\delta(t)$ such that

$$\mathcal{L}\{\delta(t)\} = 1.$$

This means reducing the width of the top hat function so that it lies between 0 and $1/T$ (not $-1/T$ and $1/T$) and increasing the height from $\frac{1}{2}T$ to T in order to preserve unit area. Clearly the difficulty arises because the impulse function is centred on $t = 0$ which is precisely the lower limit of the integral in the definition of the Laplace Transform. Using $0-$ as the lower limit of the integral overcomes many of the difficulties.

The function $\delta(t - t_0)$ represents an impulse that is centred on the time $t = t_0$. It can be considered to be the limit of the function $K(t)$ where $K(t)$ is the displaced top hat function defined by

$$K(t) = \begin{cases} 0 & t \le t_0 - 1/2T \\ \frac{1}{2}T & t_0 - 1/2T < t < t_0 + 1/2T \\ 0 & t \ge t_0 + 1/2T \end{cases}$$

as $T \to \infty$. The definition of the delta function can be used to deduce that

$$\int_{-\infty}^{\infty} h(t)\delta(t - t_0)dt = h(t_0)$$

and that, provided $t_0 > 0$

$$\mathcal{L}\{\delta(t - t_0)\} = e^{-st_0}.$$

Letting $t_0 \to 0$ leads to

$$\mathcal{L}\{\delta(t)\} = 1$$

a correct result. Another interesting result can be deduced almost at once and expresses mathematically the property of $\delta(t)$ to pick out a particular function value, known to engineers as the *filtering property*. Since

$$\int_{-\infty}^{\infty} h(t)\delta(t - t_0)dt = h(t_0)$$

with $h(t) = e^{-st}f(t)$ and $t_0 = a \ge 0$ we deduce that

$$\mathcal{L}\{\delta(t - a)f(t)\} = e^{-as}f(a).$$

Mathematically, the impulse function has additional interest in that it enables insight to be gained into the properties of discontinuous functions. From a practical point of view too there are a number of real phenomena that are closely approximated by the delta function. The sharp blow from a hammer, the discharge of a capacitor or even the sound of the bark of a dog are all in some sense impulses. All of this provides motivation for the study of the delta function.

One property that is particularly useful in the context of Laplace Transforms is the value of the integral

$$\int_{-\infty}^{t} \delta(u - u_0)du.$$

This has the value 0 if $u_0 > t$ and the value 1 if $u_0 < t$. Thus we can write

$$\int_{-\infty}^{t} \delta(u - u_0)du = \begin{cases} 0 & t < u_0 \\ 1 & t > u_0 \end{cases}$$

or

$$\int_{-\infty}^{t} \delta(u - u_0) du = H(t - u_0)$$

where H is Heaviside's Unit Step Function. If we were allowed to differentiate this result, or to put it more formally to use the fundamental theorem of the calculus (on functions one of which is not really a function, a second which is not even continuous let alone differentiable!) then one could write that "$\delta(u - u_0) = H'(u - u_0)$" or state that "the impulse function is the derivative of the Heaviside Unit Step Function". Before the pure mathematicians send out lynching parties, let us examine these loose notions. Everywhere except where $u = u_0$ the statement is equivalent to stating that the derivative of unity is zero, which is obviously true. The additional information in the albeit loose statement in quotation marks is a quantification of the nature of the unit jump in $H(u - u_0)$. We know the gradient there is infinite, but the nature of it is embodied in the second integral condition in the definition of the delta function, Equation 2.4. The subject of *generalised functions* is introduced through this concept and the interested reader is directed towards the texts by Jones and Lighthill. There will be some further detail in this book, but not until Chapter 6. All that will be noted here is that it is possible to define a whole string of derivatives $\delta'(t)$, $\delta''(t)$, etc. where all these derivatives are zero everywhere except at $t = 0$. The key to keeping rigorous here is the property

$$\int_{-\infty}^{\infty} h(t)\delta(t) dt = h(0).$$

The "derivatives" have analogous properties, viz.

$$\int_{-\infty}^{\infty} h(t)\delta'(t) dt = -h'(0)$$

and in general

$$\int_{-\infty}^{\infty} h(t)\delta^{(n)}(t) dt = (-1)^n h^{(n)}(0).$$

Of course, the function $h(t)$ will have to be appropriately differentiable. In this chapter, the Laplace Transform of this nth derivative of the Dirac delta function is required. It can be easily deduced that

$$\int_{-\infty}^{\infty} e^{-st} \delta^{(n)}(t) dt = \int_{0-}^{\infty} e^{-st} \delta^{(n)}(t) dt = s^n.$$

Notice that for all these generalised functions, the condition for the validity of the initial value theorem is violated, and the final value theorem although perfectly valid is entirely useless. It is time to do a few examples.

Example 2.16 *Determine the Inverse Laplace Transform*

$$\mathcal{L}^{-1}\left\{ \frac{s^2}{s^2 + 1} \right\}$$

and interpret the $F(t)$ obtained.

Solution Writing

$$\frac{s^2}{s^2+1} = 1 - \frac{1}{s^2+1}$$

and using the linearity property of the Inverse Laplace Transform gives

$$\mathcal{L}^{-1}\left\{\frac{s^2}{s^2+1}\right\} = \mathcal{L}^{-1}\{1\} + \mathcal{L}^{-1}\left\{\frac{1}{s^2+1}\right\}$$

$$= \delta(t) - \sin(t).$$

This function is sinusoidal with a unit impulse at $t = 0$.

Note the direct use of the inverse $\mathcal{L}^{-1}\{1\} = \delta(t)$. This arises straight away from our definition of $\mathcal{L}$. It is quite possible for other definitions of Laplace Transform to give the value $\frac{1}{2}$ for $\mathcal{L}\{\delta(t)\}$ (for example). This may worry those readers of a pure mathematical bent. However, as long as there is consistency in the definitions of the delta function and the Laplace Transform and hence its inverse, then no inconsistencies arise. The example given above will always yield the same answer $\mathcal{L}^{-1}\left\{\dfrac{s^2}{s^2+1}\right\} = \delta(t) - \sin(t)$. The small variations possible in the definition of the Laplace Transform around $t = 0$ do not change this. Our definition, viz.

$$\mathcal{L}\{F(t)\} = \int_{0-}^{\infty} e^{-st} F(t) dt$$

remains the most usual.

Example 2.17 *Find the value of* $\mathcal{L}^{-1}\left\{\dfrac{s^3}{s^2+1}\right\}$.

Solution Using a similar technique to the previous example we first see that

$$\frac{s^3}{s^2+1} = s - \frac{s}{s^2+1}$$

so taking Inverse Laplace Transforms using the linearity property once more yields

$$\mathcal{L}^{-1}\left\{\frac{s^3}{s^2+1}\right\} = \mathcal{L}^{-1}\{s\} + \mathcal{L}^{-1}\left\{\frac{s}{s^2+1}\right\}$$

$$= \delta'(t) - \cos(t)$$

where $\delta'(t)$ is the first derivative of the Dirac-δ function which was defined earlier.

Notice that the first derivative formula:

$$\mathcal{L}\{F'(t)\} = sf(s) - F(0)$$

with $F'(t) = \delta'(t) - \cos(t)$ gives

$$\mathcal{L}\{\delta'(t) - \cos(t)\} = \frac{s^3}{s^2+1} - F(0)$$

which is indeed the above result apart from the troublesome $F(0)$. $F(0)$ is of course not defined! Care indeed is required if standard Laplace Transform results are to be applied to problems containing generalised functions. When in doubt, the best advice is to use limit definitions of $\delta(t)$ and the like, and follow the mathematics through carefully, especially the swapping of integrals and limits. The little book by Lighthill is full of excellent practical advice.

2.7   Periodic Functions

We begin with a very straightforward definition that should be familiar to everyone:

Definition 2.18 *If $F(t)$ is a function that obeys the rule*

$$F(t) = F(t + \tau)$$

for some real τ for all values of t then $F(t)$ is called a periodic function with period τ.

Periodic functions play a very important role in many branches of engineering and applied science, particularly physics. One only has to think of springs or alternating current present in household electricity to realise their prevalence. In Chapter 4, we study Fourier series in some depth. This is a systematic study of periodic functions. Here, a theorem on the Laplace Transform of periodic functions is introduced, proved and used in some illustrative examples.

Theorem 2.19 *Let $F(t)$ have period $T > 0$ so that $F(t) = F(t + T)$. Then*

$$\mathcal{L}\{F(t)\} = \frac{\int_0^T e^{-st} F(t) dt}{1 - e^{-sT}}.$$

Proof Like many proofs of properties of Laplace Transforms, this one begins with its definition then evaluates the integral by using the periodicity of $F(t)$

$$\mathcal{L}\{F(t)\} = \int_0^\infty e^{-st} F(t) dt$$

$$= \int_0^T e^{-st} F(t) dt + \int_T^{2T} e^{-st} F(t) dt$$

$$+ \int_{2T}^{3T} e^{-st} F(t) dt + \cdots + \int_{(n-1)T}^{nT} e^{-st} F(t) dt + \cdots$$

provided the series on the right hand side is convergent. This is assured since the function $F(t)$ satisfies the condition for the existence of its Laplace Transform by construction. Consider the integral

$$\int_{(n-1)T}^{nT} e^{-st} F(t) dt$$

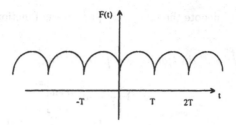

Figure 2.5: The graph of $F(t)$

and substitute $u = t - (n-1)T$. Since F has period T this leads to

$$\int_{(n-1)T}^{nT} e^{-st} F(t)dt = e^{-s(n-1)T} \int_{0}^{T} e^{-su} F(u)du \quad n = 1, 2, \ldots$$

which gives

$$\int_{0}^{\infty} e^{-st} F(t)dt = (1 + e^{-sT} + e^{-2sT} + \cdots) \int_{0}^{T} e^{-st} F(t)dt$$

$$= \frac{\int_{0}^{T} e^{-st} F(t)dt}{1 - e^{-sT}}$$

on summing the geometric progression. This proves the result.

□

Here is an example of using this theorem.

Example 2.20 *A rectified sine wave is defined by the expression*

$$F(t) = \begin{cases} \sin(t) & 0 < t < \pi \\ -\sin(t) & \pi < t < 2\pi \end{cases}$$

$$F(t) = F(t + 2\pi)$$

determine $L\{F(t)\}$.

Solution The graph of $F(t)$ is shown in Figure 2.5. The function $F(t)$ actually has period π, but it is easier to carry out the calculation as if the period was 2π. Additionally we can check the answer by using the theorem with $T = \pi$. With $T = 2\pi$ we have from Theorem 2.19,

$$\mathcal{L}\{F(t)\} = \frac{\int_{0}^{2\pi} e^{-st} F(t)dt}{1 - e^{-sT}}$$

where the integral in the numerator is evaluated by splitting into two as follows:-

$$\int_{0}^{2\pi} e^{-st} F(t)dt = \int_{0}^{\pi} e^{-st} \sin(t)dt + \int_{\pi}^{2\pi} e^{-st}(-\sin(t))dt.$$

Now, writing $\Im\{\}$ to denote the imaginary part of the function in the brace we have

$$\int_0^\pi e^{-st}\sin(t)dt = \Im\left\{\int_0^\pi e^{-st+it}dt\right\}$$

$$= \Im\left[\frac{1}{i-s}e^{-st+it}\right]_0^\pi$$

$$= \Im\left\{\frac{1}{i-s}(e^{-s\pi+i\pi}-1)\right\}$$

$$= \Im\left\{\frac{1}{s-i}(1+e^{-s\pi})\right\}.$$

So

$$\int_0^\pi e^{-st}\sin(t)dt = \frac{1+e^{-\pi s}}{1+s^2}.$$

Similarly,

$$\int_\pi^{2\pi} e^{-st}\sin(t)dt = -\frac{e^{-2\pi s}+e^{-\pi s}}{1+s^2}.$$

Hence we deduce that

$$\mathcal{L}\{F(t)\} = \frac{(1+e^{-\pi s})^2}{(1+s^2)(1-e^{-2\pi s})}$$

$$= \frac{1+e^{-\pi s}}{(1+s^2)(1-e^{-\pi s})}.$$

This is precisely the answer that would have been obtained if Theorem 2.19 had been applied to the function

$$F(t) = \sin(t)\ \ 0 < t < \pi\ \ \ F(t) = F(t+\pi).$$

We can therefore have some confidence in our answer.

2.8   Exercises

1. If $F(t) = \cos(at)$, use the derivative formula to re-establish the Laplace Transform of $\sin(at)$.

2. Use Theorem 2.2 with

$$F(t) = \int_0^t \frac{\sin(u)}{u}du$$

to establish the result.

$$\mathcal{L}\left\{\frac{\sin(at)}{t}\right\} = \tan^{-1}\left\{\frac{a}{s}\right\}.$$

3. Prove that

$$\mathcal{L}\left\{ \int_0^t \int_0^v F(u)du\,dv \right\} = \frac{f(s)}{s^2}.$$

4. Find

$$\mathcal{L}\left\{ \int_0^t \frac{\cos(au) - \cos(bu)}{u} du \right\}.$$

5. Determine

$$\mathcal{L}\left\{ \frac{2\sin(t)\sinh(t)}{t} \right\}.$$

6. Prove that if $\bar{f}(s)$ indicates the Laplace Transform of a piecewise continuous function $f(t)$ then

$$\lim_{s \to \infty} \bar{f}(s) = 0.$$

7. Determine the following Inverse Laplace Transforms by using partial fractions

(a) $\dfrac{2(2s + 7)}{(s + 4)(s + 2)}$, $s > -2$ (b) $\dfrac{s + 9}{s^2 - 9}$,

(c) $\dfrac{s^2 + 2k^2}{s(s^2 + 4k^2)}$, (d) $\dfrac{1}{s(s + 3)^2}$,

(e) $\dfrac{1}{(s - 2)^2(s + 3)^3}$.

8. Verify the initial value theorem, Theorem 2.12, for the two functions
(a) $2 + \cos(t)$ and
(b) $(4 + t)^2$.

9. Verify the final value theorem, Theorem 2.13, for the two functions
(a) $3 + e^{-t}$ and
(b) $t^3 e^{-t}$.

10. Given that

$$\mathcal{L}\{\sin(\sqrt{t})\} = \frac{k}{s^{3/2}} e^{-1/4s}$$

use $\sin x \sim x$ near $x = 0$ to determine the value of the constant k. (You will need the table of standard transforms Appendix B.)

11. By using a power series expansion, determine (in series form) the Laplace Transforms of $\sin(t^2)$ and $\cos(t^2)$.

12. $P(s)$ and $Q(s)$ are polynomials, the degree of $P(s)$ is less than that of $Q(s)$ which is n. Use partial fractions to prove the result

$$\mathcal{L}^{-1}\left\{ \frac{P(s)}{Q(s)} \right\} = \sum_{k=1}^{n} \frac{P(\alpha_k)}{Q'(\alpha_k)} e^{\alpha_k t}$$

where α_k are the n distinct zeros of $Q(s)$.

13. Find the following Laplace Transforms:
    (a) $H(t - a)$
    (b)

$$f_1 = \begin{cases} t+1 & 0 \le t \le 2 \\ 3 & t > 2 \end{cases}$$

    (c)

$$f_2 = \begin{cases} t+1 & 0 \le t \le 2 \\ 6 & t > 2 \end{cases}$$

    (d) the derivative of $f_1(t)$.

14. Find the Laplace Transform of the triangular wave function:

$$F(t) = \begin{cases} t & 0 \le t < c \\ 2c - t & c \le t < 2c \end{cases}$$
$$F(t + 2c) = F(t).$$

3

Convolution and the Solution of Ordinary Differential Equations

3.1 Introduction

It is assumed from the outset that students will have some familiarity with ordinary differential equations (ODE), but there is a brief résumé given in Section 3.3. The other central and probably new idea is that of the convolution integral and this is introduced fully in Section 3.2. Of course it is possible to solve some kinds of differential equation without using convolution as is obvious from the last chapter, but mastery of the convolution theorem greatly extends the power of Laplace Transforms to solve ODEs. In fact, familiarity with the convolution operation is necessary for the understanding of many other topics that feature in this text such as the solution of partial differential equations (PDEs) and other topics that are outside it such as the use of Green's functions for forming the general solution of various types of boundary value problem (BVP).

3.2 Convolution

The definition of convolution is straightforward.

Definition 3.1 *The convolution of two given functions $f(t)$ and $g(t)$ is written $f * g$ and is defined by the integral*

$$f * g = \int_0^t f(\tau)g(t - \tau)d\tau.$$

The only condition that is necessary to impose on the functions f and g is that their behaviour be such that the integral on the right exists. Piecewise continuity of both in the interval $[0, t]$ is certainly sufficient. The following definition of piecewise continuity is repeated here for convenience.

Definition 3.2 *If an interval* $[0, t_0]$ *say can be partitioned into a finite number of subintervals* $[0, t_1], [t_1, t_2], [t_2, t_3], \ldots, [t_n, t_0]$ *with* $0, t_1, t_2, \ldots, t_n, t_0$ *an increasing sequence of times and such that a given function* $f(t)$ *is continuous in each of these subintervals but not necessarily at the end points themselves, then* $f(t)$ *is piecewise continuous in the interval* $[0, t_0]$.

It is easy to prove the following theorem

Theorem 3.3 (Symmetry) $f * g = g * f$.

It is left as an exercise to the student to prove this.

Probably the most important theorem concerning the use of Laplace Transforms and convolution is introduced now. It is called the convolution theorem and enables one, amongst other things, to deduce the Inverse Laplace Transform of an expression provided the expression can be expressed in the form of a product of functions, each Inverse Laplace Transform of which is known. Thus, in a loose sense, the Inverse Laplace Transform is equivalent to integration by parts, although unlike integration by parts, there is no integration left to do on the right hand side.

Theorem 3.4 (Convolution) *If* $f(t)$ *and* $g(t)$ *are two functions of exponential order (so that their Laplace Transforms exist – see Chapter 1), and writing* $\mathcal{L}\{f\} = \bar{f}(s)$ *and* $\mathcal{L}\{g\} = \bar{g}(s)$ *as the two Laplace Transforms then* $\mathcal{L}^{-1}\{\bar{f}\bar{g}\} = f * g$ *where* $*$ *is the convolution operator introduced above.*

Proof In order to prove this theorem, we in fact show that

$$\bar{f}\bar{g} = \mathcal{L}\{f(t) * g(t)\}$$

by direct integration of the right hand side. In turn, this involves its interpretation in terms of a repeated integral. Now,

$$\mathcal{L}\{f(t) * g(t)\} = \int_0^\infty e^{-st} \int_0^t f(\tau)g(t - \tau)d\tau dt$$

using the definition of the Laplace Transform. The domain of this repeated integral takes the form of a wedge in the t, τ plane. This wedge (infinite wedge) is displayed in Figure 3.1. Trivial rewriting of this double integral to facilitate changing the order of integration gives

$$\mathcal{L}\{f(t) * g(t)\} = \int_0^\infty \int_0^t e^{-st} f(\tau)g(t - \tau)d\tau dt$$

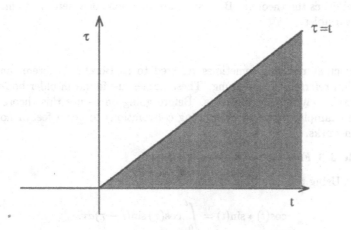

Figure 3.1: The domain of the repeated integral

and thus integrating with respect to t first (horizontally first instead of vertically in Figure 3.1) gives

$$\mathcal{L}\{f(t) * g(t)\} = \int_0^\infty \int_\tau^\infty e^{-st} f(\tau) g(t - \tau) dt d\tau$$

$$= \int_0^\infty f(\tau) \left\{ \int_\tau^\infty e^{-st} g(t - \tau) dt \right\} d\tau.$$

Implement the change of variable $u = t - \tau$ in the inner integral where τ is constant so that it becomes

$$\int_\tau^\infty e^{-st} g(t - \tau) dt = \int_0^\infty e^{-s(u+\tau)} g(u) du$$

$$= e^{-s\tau} \int_0^\infty e^{-su} g(u) du$$

$$= e^{-s\tau} \bar{g}(s).$$

Thus we have

$$\mathcal{L}\{f * g\} = \int_0^\infty f(\tau) e^{-s\tau} \bar{g}(s) d\tau$$

$$= \bar{g}(s) \bar{f}(s)$$

$$= \bar{f}(s) \bar{g}(s).$$

Hence

$$f(t) * g(t) = \mathcal{L}^{-1}\{\bar{f}\bar{g}\}.$$

This establishes the theorem. Both sides are functions of s here, τ, t and u are dummy variables.

□

This particular result is sometimes referred to as Borel's Theorem, and the convolution referred to as Faltung. These names are found in older books and some present day engineering texts. Before going on to use this theorem, let us do an example or two on calculating convolutions to get a feel of how the operation works.

Example 3.5 *Find the value of* $\cos(t) * \sin(t)$.

Solution Using the definition of convolution we have

$$\cos(t) * \sin(t) = \int_0^t \cos(\tau)\sin(t - \tau)d\tau.$$

To evaluate this we could of course resort to computer algebra: alternatively we use the identity

$$\sin(A)\cos(B) = \frac{1}{2}[\sin(A + B) + \sin(A - B)].$$

This identity is engraved on the brains of those who passed exams before the advent of formula sheets. It does however appear on most trigonometric formula sheets. Evaluating the integral by hand thus progresses as follows. Let $A = t - \tau$ and $B = \tau$ in this trigonometric formula to obtain

$$\sin(t - \tau)\cos(\tau) = \frac{1}{2}[\sin(t) + \sin(t - 2\tau)]$$

whence,

$$\begin{aligned}
\cos(t) * \sin(t) &= \int_0^t \cos(\tau)\sin(t - \tau)d\tau \\
&= \frac{1}{2}\int_0^t [\sin(t) + \sin(t - 2\tau)]d\tau \\
&= \frac{1}{2}\sin(t)[\tau]_0^t + \frac{1}{4}[\cos(t - 2\tau)]_0^t \\
&= \frac{1}{2}t\sin(t) + \frac{1}{4}[\cos(-t) - \cos(t)] \\
&= \frac{1}{2}t\sin(t).
\end{aligned}$$

Let us try another example of evaluating a convolution integral.

Example 3.6 *Find the value of* $\sin(t) * t^2$.

Solution We progress as before by using the definition

$$\sin(t) * t^2 = \int_0^t \sin(\tau)(t - \tau)^2 d\tau.$$

It is up to us to choose the order as from Theorem 3.3 $f*g = g*f$. Of course we choose the order that gives the easier integral to evaluate. In fact there is little to choose in this present example, but it is a point worth watching in future. This integral is evaluated by integration by parts. Here are the details.

$$\begin{aligned}
\sin(t) * t^2 &= \int_0^t \sin(\tau)(t - \tau)^2 d\tau \\
&= [-(t - \tau)^2 \cos(\tau)]_0^t - \int_0^t 2(t - \tau) \cos(\tau) d\tau \\
&= t^2 - 2\left\{ [(t - \tau) \sin(\tau)]_0^t - \int_0^t \sin(\tau) d\tau \right\} \\
&= t^2 - 2\left\{ 0 + [-\cos(\tau)]_0^t \right\} \\
&= t^2 + 2\cos(t) - 2.
\end{aligned}$$

Of course the integration can be done by computer algebra.

Both of these examples provide typical evaluations of convolution integrals. Convolution integrals occur in many different branches of engineering, particularly when signals are being processed (see Section 6.5). However, we shall only be concerned with their application to the evaluation of the Inverse Laplace Transform. Therefore without further ado, let us do a couple of these examples.

Example 3.7 *Find the following Inverse Laplace Transforms:*

$$(a)\ \mathcal{L}^{-1}\left\{ \frac{s}{(s^2 + 1)^2} \right\},$$

$$(b)\ \mathcal{L}^{-1}\left\{ \frac{1}{s^3(s^2 + 1)} \right\}.$$

Solution (a) We cannot evaluate this Inverse Laplace Transform in any direct fashion. However we do know the standard forms

$$\mathcal{L}^{-1}\left\{ \frac{s}{(s^2 + 1)} \right\} = \cos(t) \text{ and } \mathcal{L}^{-1}\left\{ \frac{1}{(s^2 + 1)} \right\} = \sin(t).$$

Hence

$$\mathcal{L}\{\cos(t)\}\mathcal{L}\{\sin(t)\} = \frac{s}{(s^2 + 1)^2}$$

and so using the convolution theorem, and Example 3.1

$$\mathcal{L}^{-1}\left\{ \frac{s}{(s^2 + 1)^2} \right\} = \cos(t) * \sin(t) = \frac{1}{2}t \sin(t).$$

(b) Proceeding similarly with this Inverse Laplace Transform, we identify the standard forms:-

$$\mathcal{L}^{-1}\left\{\frac{1}{s^3}\right\} = \frac{1}{2}t^2 \text{ and } \mathcal{L}^{-1}\left\{\frac{1}{(s^2+1)}\right\} = \sin(t).$$

Thus

$$\mathcal{L}\{t^2\}\mathcal{L}\{\sin(t)\} = \frac{2}{s^3(s^2+1)}$$

and so

$$\mathcal{L}^{-1}\left\{\frac{1}{s^3(s^2+1)}\right\} = \frac{1}{2}t^2 * \sin(t).$$

This convolution has been found in Example 3.6, hence the Inverse Laplace Transform is

$$\mathcal{L}^{-1}\left\{\frac{1}{s^3(s^2+1)}\right\} = \frac{1}{2}(t^2 + 2\cos t - 2).$$

In this case there is an alternative approach as the expression

$$\frac{1}{s^3(s^2+1)}$$

can be decomposed into partial fractions and the inverse Laplace Transform evaluated by the methods of Chapter 2.

In a sense, the last example was "cooked" in that the required convolutions just happened to be those already evaluated. Nevertheless the power of the convolution integral is clearly demonstrated. In the kind of examples met here there is usually little doubt that the functions meet the conditions necessary for the existence of the Laplace Transform. In the real world, the functions may be time series, have discontinuities or exhibit a stochastic character that makes formal checking of these conditions awkward. It remains important to do this checking however: that is the role of mathematics! Note that the product of two functions that are of exponential order is also of exponential order and the integral of this product is of exponential order too.

The next step is to use the convolution theorem on more complicated results. To do this requires the derivation of the "well known" integral

$$\int_0^\infty e^{-t^2} dt = \frac{1}{2}\sqrt{\pi}.$$

Example 3.8 *Use a suitable double integral to evaluate the improper integral*

$$\int_0^\infty e^{-t^2} dt.$$

Solution Consider the double integral

$$\int\int_S e^{-(x^2+y^2)} dS$$

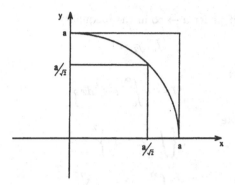

Figure 3.2: The domains of the repeated integrals

where S is the quarter disc $x \geq 0, y \geq 0$, $x^2 + y^2 \leq a^2$. Converting to polar co-ordinates (R, θ) this integral becomes

$$\int_0^a \int_0^{\frac{\pi}{2}} e^{-R^2} R d\theta dR$$

where $R^2 = x^2 + y^2$, $R\cos(\theta) = x$, $R\sin(\theta) = y$ so that $dS = Rd\theta dR$. Evaluating this integral we obtain

$$I = \frac{\pi}{2} \int_0^a R e^{-R^2} dR = \frac{\pi}{2} \left[-\frac{1}{2} e^{-R^2} \right]_0^a = \frac{\pi}{4} \left\{ 1 - e^{-a^2} \right\}.$$

As $a \to \infty$, $I \to \frac{\pi}{4}$.

We now consider the double integral

$$I_k = \int_0^k \int_0^k e^{-(x^2+y^2)} dx dy.$$

The domain of this integral is a square of side k. Now

$$I_k = \left\{ \int_0^k e^{-x^2} dx \right\} \left\{ \int_0^k e^{-y^2} dy \right\} = \left\{ \int_0^k e^{-x^2} dx \right\}^2$$

A glance at Figure 3.2 will show that

$$I_{a/\sqrt{2}} < I < I_a.$$

However, we can see that

$$I_k \to \left\{ \int_0^\infty e^{-x^2} dx \right\}^2$$

as $k \to \infty$. Hence if we let $a \to \infty$ in the inequality

$$I_{a/\sqrt{2}} < I < I_a$$

we deduce that

$$I \to \left\{ \int_0^\infty e^{-x^2} dx \right\}^2$$

as $a \to \infty$. Therefore

$$\left\{ \int_0^\infty e^{-x^2} dx \right\}^2 = \frac{\pi}{4}$$

or

$$\int_0^\infty e^{-x^2} dx = \frac{\sqrt{\pi}}{2}$$

as required. This formula plays a central role in statistics, being one half of the area under the bell shaped curve usually associated with the normal distribution. In this text, its frequent appearance in solutions to problems involving diffusion and conduction is more relevant.

Let us use this integral to find the Laplace Transform of $1/\sqrt{t}$. From the definition of Laplace Transform, it is necessary to evaluate the integral

$$\mathcal{L} \left\{ \frac{1}{\sqrt{t}} \right\} = \int_0^\infty \frac{e^{-st}}{\sqrt{t}} dt$$

provided we can be sure it exists. The behaviour of the integrand at ∞ is not in question, and the potential problem at the origin disappears once it is realised that $1/\sqrt{t}$ itself is integrable in any finite interval that contains the origin. We can therefore proceed to evaluate it. This is achieved through the substitution $st = u^2$. The conversion to the variable u results in

$$\int_0^\infty \frac{e^{-st}}{\sqrt{t}} dt = \int_0^\infty \frac{e^{-u^2}}{\frac{\sqrt{s}}{}} \frac{\sqrt{s}}{u} 2u \, du = \frac{2}{\sqrt{s}} \int_0^\infty e^{-u^2} du = \sqrt{\frac{\pi}{s}}.$$

We have therefore established the Laplace Transform result:

$$\mathcal{L} \left\{ \frac{1}{\sqrt{t}} \right\} = \sqrt{\frac{\pi}{s}}$$

and, perhaps more importantly, the inverse

$$\mathcal{L}^{-1} \left\{ \frac{1}{\sqrt{s}} \right\} = \frac{1}{\sqrt{\pi t}}.$$

These results have wide application. Their use with the convolution theorem opens up a whole new class of functions on which the Laplace Transform and its inverse can operate. The next example is typical.

Example 3.9 *Determine*

$$\mathcal{L}^{-1} \left\{ \frac{1}{\sqrt{s}(s-1)} \right\}.$$

Solution The only sensible way to proceed using our present knowledge (but see Chapter 7) is to use the results

$$\mathcal{L}^{-1}\left\{\frac{1}{\sqrt{s}}\right\} = \frac{1}{\sqrt{\pi t}}$$

and (by the shifting property of Chapter 2)

$$\mathcal{L}^{-1}\left\{\frac{1}{s-1}\right\} = e^t.$$

Whence, using the convolution theorem

$$\mathcal{L}^{-1}\left\{\frac{1}{\sqrt{s}(s-1)}\right\} = \int_0^t \frac{1}{\sqrt{\pi\tau}} e^{(t-\tau)} d\tau$$

$$= \frac{e^t}{\sqrt{\pi}} \int_0^t \frac{e^{-\tau}}{\sqrt{\tau}} d\tau.$$

The last integral is similar to that evaluated in determining the Laplace Transform of $1/\sqrt{t}$ except for the upper limit. It is tackled using the same transformation, only this time we keep careful track of the upper limit. Therefore, substitute $\tau = u^2$ to obtain

$$\int_0^t \frac{e^{-\tau}}{\sqrt{\tau}} d\tau = 2 \int_0^{\sqrt{t}} e^{-u^2} du.$$

Now, this integral is well known from the definition of the Error Function, erf(x).

Definition 3.10 *The Error Function erf(x) is defined by*

$$erf(x) = \frac{2}{\sqrt{\pi}} \int_0^x e^{-t^2} dt.$$

It is related to the area under the normal distribution curve in statistics. The factor of $2/\sqrt{\pi}$ is there to ensure that the total area is unity as is required by probability theory. Returning to Laplace Transforms, it is thus possible to express the solution as follows:-

$$\mathcal{L}^{-1}\left\{\frac{1}{\sqrt{s}(s-1)}\right\} = e^t erf\{\sqrt{t}\}.$$

The function $1-erf(x)$ is called the complementary error function and is written erfc(x). It follows immediately that

$$erfc(x) = \frac{2}{\sqrt{\pi}} \int_x^\infty e^{-t^2} dt \text{ and } erf(x) + erfc(x) = 1.$$

It is in fact the complementary error function rather than the error function itself that emerges from solving the differential equations related to diffusion problems. Having solved one simple example, it is appropriate to invert the Laplace Transform that occurs time and time again in the solution of diffusion problems. These problems are met explicitly in Chapter 5, but here we solve the problem of finding the Laplace Transform of the function

$$t^{-3/2} \exp\left\{-\frac{k^2}{4t}\right\}$$

where k is a constant representing the diffusion called the diffusion coefficient. Here it can be thought of as an unspecified constant.

Example 3.11 *Determine*

$$\mathcal{L}\left\{t^{-3/2} \exp\left\{-\frac{k^2}{4t}\right\}\right\}.$$

Solution We start as always by writing the Laplace Transform explicitly in terms of the integral definition, and it looks daunting. To evaluate it does require some convoluted algebra. As with all algebraic manipulation these days, the option is there to use computer algebra, although existing systems would find this hard. In any case, the derivation of this particular formula by hand does help in the understanding of its subsequent use. So, we start with

$$\mathcal{L}\left\{t^{-3/2} \exp\left\{-\frac{k^2}{4t}\right\}\right\} = \int_0^\infty e^{-k^2/4t} e^{-st} t^{-3/2} dt.$$

First of all, let us substitute $u = k/2\sqrt{t}$. This is done so that a term e^{-u^2} appears in the integrand. Much other stuff appears too of course and we now sort this out:

$$du = -\frac{k}{4} t^{-3/2} dt$$

which eliminates the $t^{-3/2}$ term. The limits swap round, then swap back again once the negative sign from the du term is taken into account. Thus we obtain

$$\int_0^\infty e^{-k^2/4t} e^{-st} t^{-3/2} dt = \frac{4}{k} \int_0^\infty e^{-u^2} e^{-sk/4u^2} du$$

and this completes the first stage. We now strive to get the integral on the right into error function form. (Computer algebra systems need leading by the nose through this kind of algebra!) First of all we complete the square

$$u^2 + \frac{sk^2}{4u^2} = \left\{u - \frac{k\sqrt{s}}{2u}\right\}^2 + k\sqrt{s}.$$

The unsquared term is independent of u; this is important. The integral can thus be written:

$$\int_0^\infty e^{-k^2/4t} e^{-st} t^{-3/2} dt = \frac{4}{k} e^{-k\sqrt{s}} \int_0^\infty e^{-\left(u - \frac{k\sqrt{s}}{2u}\right)^2} du.$$

This completes the second stage. The third and perhaps most bizarre stage of evaluating this integral involves consideration and manipulation of the integral

$$\int_0^\infty e^{-(u-\frac{a}{u})^2}\,du, \text{ where } a = \frac{k\sqrt{s}}{2}.$$

If we let $v = a/u$ in this integral, then since

$$\left(u - \frac{a}{u}\right)^2 = \left(v - \frac{a}{v}\right)^2$$

but $du = -adv/v^2$ we obtain the unexpected result

$$\int_0^\infty e^{-(u-\frac{a}{u})^2}\,du = \int_0^\infty \frac{a}{u^2} e^{-(u-\frac{a}{u})^2}\,du.$$

The minus sign cancels with the exchange in limits as before, and the dummy variable v has been replaced by u. We can use this result to deduce immediately that

$$\int_0^\infty \left(1 + \frac{a}{u^2}\right) e^{-(u-\frac{a}{u})^2}\,du = 2\int_0^\infty e^{-(u-\frac{a}{u})^2}\,du.$$

In the left hand integral, we substitute $\lambda = u - a/u$ so that $d\lambda = (1 + a/u^2)du$. In this way, we regain our friend from Example 3.8, apart from the lower limit which is $-\infty$ rather than 0. Finally therefore

$$\int_0^\infty \left(1 + \frac{u}{u^2}\right) e^{-(u-\frac{a}{u})^2}\,du = \int_{-\infty}^\infty e^{-\lambda^2}\,d\lambda$$
$$= 2\int_0^\infty e^{-\lambda^2}\,d\lambda$$
$$= \sqrt{\pi}.$$

Hence we have deduced that

$$\int_0^\infty e^{-(u-\frac{a}{u})^2}\,du = \frac{1}{2}\sqrt{\pi}$$

and is independent of the constant a. Using these results, a summary of the calculation of the required Laplace Transform is

$$\mathcal{L}\left\{t^{-3/2}\exp\left\{-\frac{k^2}{4t}\right\}\right\} = \frac{4}{k}e^{-k\sqrt{s}}\int_0^\infty e^{(u-\frac{k\sqrt{s}}{2u})^2}\,du.$$
$$= \frac{4}{k}e^{-k\sqrt{s}}\frac{1}{2}\sqrt{\pi}$$
$$= \frac{2\sqrt{\pi}}{k}e^{-k\sqrt{s}}.$$

Taking the Inverse Laplace Transform of this result gives the equally useful formula

$$\mathcal{L}^{-1}\left\{e^{-k\sqrt{s}}\right\} = \frac{k}{2\sqrt{\pi t^3}}e^{-k^2/4t}.$$

As mentioned earlier, this Laplace Transform occurs in diffusion and conduction problems. In particular for the applied mathematician, it enables the estimation of possible time scales for the diffusion of pollutant from a point source. Let us do one more example using the result just derived.

Example 3.12 *Use the convolution theorem to find*

$$\mathcal{L}^{-1}\left\{\frac{e^{-k\sqrt{s}}}{s}\right\}.$$

Solution We note the result just derived, namely

$$\mathcal{L}^{-1}\left\{e^{-k\sqrt{s}}\right\} = \frac{k}{2\sqrt{\pi t^3}}e^{-k^2/4t}.$$

together with the standard result

$$\mathcal{L}^{-1}\left\{\frac{1}{s}\right\} = 1$$

to deduce that

$$\mathcal{L}^{-1}\left\{\frac{e^{-k\sqrt{s}}}{s}\right\} = \frac{k}{2\sqrt{\pi t^3}}e^{-k^2/4t} * 1$$

$$= \frac{k}{2\sqrt{\pi}}\int_0^t \tau^{-3/2}e^{-k^2/4\tau}d\tau.$$

We evaluate this integral by the (by now familiar) trick of substituting $u^2 = k^2/4\tau$. This means that

$$2u\,du = -\frac{k^2}{4\tau^2}d\tau$$

and the limits transform from $\tau = 0$ and $\tau = t$ to $u = \infty$ and $u = k/2\sqrt{t}$ respectively. They swap round due to the negative sign in the expression for du so we obtain

$$\frac{k}{2\sqrt{\pi}}\int_0^t \tau^{-3/2}e^{-k^2/4\tau}d\tau = \frac{2}{\sqrt{\pi}}\int_{k/2\sqrt{t}}^{\infty} e^{-u^2}du$$

$$= \text{erfc}\left(\frac{k}{2\sqrt{t}}\right).$$

Hence we have the result

$$\mathcal{L}^{-1}\left\{\frac{e^{-k\sqrt{s}}}{s}\right\} = \text{erfc}\left(\frac{k}{2\sqrt{t}}\right)$$

which is also of some significance in the modelling of diffusion. An alternative derivation of this result not using convolution is possible using a result from Chapter 2, viz.

$$\mathcal{L}^{-1}\left\{\frac{\bar{f}(s)}{s}\right\} = \int_0^t f(u)du.$$

This formula can be regarded as a special case of the convolution theorem. We shall make further use of the convolution theorem in this kind of problem in Chapter 6. In the remainder of this chapter we shall apply the results of Chapter 2 to the solution of ordinary differential equations (ODEs). This also makes use of the convolution theorem both as an alternative to using partial fractions but more importantly to enable general solutions to be written down explicitly even where the right hand side of the ODE is a general function.

3.3   Ordinary Differential Equations

At the outset we stress that all the functions in this section will be assumed to be appropriately differentiable. For the examples in this section which are algebraically explicit this is obvious, but outside this section and indeed outside this text care needs to be taken to ensure that this remains the case. It is of course a stricter criterion than that needed for the existence of the Laplace Transform (that the function be of exponential order) so using the Laplace Transform as a tool for solving ordinary differential equations is usually not a problem. On the other hand, using differential equations to establish results for Laplace Transforms is certainly to be avoided as this automatically imposes the strict condition of differentiability on the functions in them. It is perhaps the premier aim of mathematics to remove restrictions and widen the class of functions that obey theorems and results, not the other way round!

Most of you will be familiar to a greater or lesser extent with differential equations: however for completeness a résumé is now given of the basic essentials. A differential equation is an equation where the unknown is in the form of a derivative. It was seen in Chapter 2 that operating on a derivative with the Laplace Transform can eliminate the derivative, replacing each differentiation with a multiple of s. It should not be surprising therefore that Laplace Transforms are a handy tool for solving certain types of differential equation. Before going into detail, let us review some of the general terminology met in discussing differential equations.

The *order* of an ordinary differential equation is the highest derivative attained by the unknown. Thus the equation

$$\left(\frac{dy}{dx}\right)^3 + y = \sin(x)$$

is a first order equation. The equation

$$\frac{d^2y}{dx^2} + \left(\frac{dy}{dx}\right)^4 + \ln x = 0$$

is, on the other hand a second order equation. The equation

$$\left(\frac{d^3 y}{dx^3}\right)^4 + \left(\frac{dy}{dx}\right)^7 + y^8 = 0$$

is of third order. Such exotic equations will not (cannot) be solved using Laplace Transforms. Instead we shall be restricted to solving first and second order equations which are linear. Linearity has been met in Chapter 1 in the context of the Laplace Transform. The linearity property is defined by

$$L\{\alpha y_1 + \beta y_2\} = \alpha L\{y_1\} + \beta L\{y_2\}.$$

In a linear differential equation, the dependent variable obeys this linearity property. We shall only be considering linear differential equations here (Laplace Transforms being linear themselves are only useful for solving linear differential equations). Differential equations that *cannot* be solved are those containing powers of the unknown or expressions such as $\tan(y)$, e^y. Thus we will solve first and second order linear differential equations. Although this seems rather restrictive, it does account for nearly all those linear ODEs found in real life situations. The word *ordinary* denotes that there is only differentiation with respect to a single independent variable so that the solution is the required function of this variable. If the number of variables is two or more, the differential equation becomes a *partial* differential equation (PDE) and these are considered later in Chapter 5.

The Laplace Transform of a derivative was found easily by direct integration by parts in Chapter 2. The two useful results that will be used extensively here are

$$\mathcal{L}\{f'(t)\} = s\bar{f}(s) - f(0)$$

and

$$\mathcal{L}\{f''(t)\} = s^2 \bar{f}(s) - sf(0) - f'(0)$$

where the prime denotes differentiation with respect to t. Note that the right hand sides of both of these expressions involve knowledge of $f(t)$ at $t = 0$. This is important. Also, the order of the derivative determines how many arbitrary constants the solution contains. There is one arbitrary constant for each integration, so a first order ODE will have one arbitrary constant, a second order ODE two arbitrary constants, etc. There are complications over uniqueness with differential equations that are not linear, but fortunately this does not concern us here. We know from Theorem 1.2 that the Laplace Transform is a linear operator; it is not easily applied to non-linear problems.

Upon taking Laplace Transforms of a linear ODE, the derivatives themselves disappear, transforming into the Laplace Transform of the function multiplied by s (for a first derivative) or s^2 (for a second derivative). Moreover, the correct number of constants also appear in the form of $f(0)$ (for a first order ODE) and $f(0)$ and $f'(0)$ (for a second order ODE). Some texts conclude therefore that Laplace Transforms can be used only to solve *initial value problems*, that is problems where enough information is known at the start to solve it. This is not strictly true. Whilst it remains the case that initial value problems are best suited to this method of solution, two point boundary value problems can be solved by transforming the equation, and retaining $f(0)$ and $f'(0)$ (or the first only) as unknowns. These unknowns are then found algebraically by the substitution of the given boundary conditions and the solving of the resulting differential equation. We shall, however almost always be solving ODEs with initial conditions (but see Example 3.14). From a physical standpoint this is entirely reasonable. The Laplace Transform is a mapping from t space to s space (see Chapter 1) and t almost always corresponds to time. For problems involving time, the situation is known *now* and the equation(s) are solved in order to determine what is going on *later*. This is indeed the classical initial value problem. We are now ready to try a few examples.

Example 3.13 *Solve the first order differential equation*

$$\frac{dx}{dt} + 3x = 0 \; where \; x(0) = 1.$$

Solution Note that we have abandoned $f(t)$ for the more usual $x(t)$, but this should be regarded as a trivial change of dummy variable. This rather simple differential equation can in fact be solved by a variety of methods. Of course we use Laplace Transforms, but it is useful to check the answer by solving again using separation of variables or integrating factor methods as these will be familiar to most students. Taking Laplace transforms leads to

$$\mathcal{L}\left\{\frac{dx}{dt}\right\} + 3\mathcal{L}\{x\} = 0$$

which implies

$$s\bar{x}(s) - x(0) + 3\bar{x}(s) = 0$$

using the standard overbar to denote Laplace Transform. Since $x(0) = 1$, solving for $\bar{x}(s)$ gives

$$\bar{x}(s) = \frac{1}{s+3}$$

whence

$$x(t) = \mathcal{L}^{-1}\left\{\frac{1}{s+3}\right\} = e^{-3t}$$

using the standard form. That this is indeed the solution is easily checked.

Let us look at the same equation, but with a different boundary condition.

Example 3.14 *Solve the first order differential equation*

$$\frac{dx}{dt} + 3x = 0 \; where \; x(1) = 1.$$

Solution Proceeding as before, we now cannot insert the value of $x(0)$ so we arrive at the solution

$$x(t) = \mathcal{L}^{-1}\left\{\frac{x(0)}{s+3}\right\} = x(0)e^{-3t}.$$

We now use the boundary condition we do have to give

$$x(1) = x(0)e^{-3} = 1$$

which implies

$$x(0) = e^3$$

and the solution is

$$x(t) = e^{3(1-t)}.$$

Here is a slightly more challenging problem.

Example 3.15 *Solve the differential equation*

$$\frac{dx}{dt} + 3x = \cos 3t \ given \ x(0) = 0.$$

Solution Taking the Laplace Transform (we have already done this in Example 3.13 for the left hand side) we obtain

$$s\bar{x}(s) - x(0) + 3\bar{x}(s) = \frac{1}{s^2 + 9}$$

using standard forms. With the zero initial condition solving this for $\bar{x}(s)$ yields

$$\bar{x}(s) = \frac{1}{(s + 3)(s^2 + 9)}.$$

This solution is in the form of the product of two known Laplace Transforms. Thus we invert either using partial fractions or the convolution theorem: we choose the latter. First of all note the standard forms

$$\mathcal{L}^{-1}\left\{\frac{1}{s + 3}\right\} = e^{-3t} \ \text{and} \ \mathcal{L}^{-1}\left\{\frac{1}{s^2 + 9}\right\} = \cos(3t).$$

Using the convolution theorem yields:

$$x(t) = \mathcal{L}^{-1}\left\{\frac{1}{s + 3}\frac{1}{s^2 + 9}\right\} = \int_0^t e^{-3(t-\tau)}\cos(3\tau)d\tau.$$

(The equally valid choice of

$$\int_0^t e^{-3\tau}\cos\left(3(t - \tau)\right) d\tau$$

could have been made, but as a general rule it is better to arrange the order of the convolution so that the $(t - \tau)$ is in an exponential if you have one.) The integral is straightforward to evaluate using integration by parts or computer algebra. The gory details are omitted here. The result is

$$\int_0^t e^{3\tau}\cos(3\tau)d\tau = \frac{1}{6}(e^{3t}\cos(3t) + e^{3t}\sin(3t) - 1).$$

Thus we have

$$x(t) = e^{-3t} \int_0^t e^{3\tau} \cos(3\tau)d\tau$$

$$= \frac{1}{6}(\cos(3t) + \sin(3t)) - \frac{1}{6}e^{-3t}.$$

This solution could have been obtained by partial fractions which is algebraically simpler. It is also possible to solve the original ODE by complementary function/particular integral techniques or to use integrating factor methods. The choice is yours! In the next example there is a clear winner. It is also possible to get a closed form answer using integrating factor techniques, but using Laplace Transforms together with the convolution theorem is our choice here.

Example 3.16 *Find the general solution to the differential equation*

$$\frac{dx}{dt} + 3x = f(t) \text{ where } x(0) = 0,$$

and $f(t)$ is of exponential order (which is sufficient for the method of solution used to be valid).

Solution It is compulsory to use convolution here as the right hand side is an arbitrary function. The Laplace Transform of the equation leads directly to

$$x(t) = \mathcal{L}^{-1}\left\{\frac{\bar{f}(s)}{s+3}\right\}$$

$$= \int_0^t e^{-3(t-\tau)}f(\tau)d\tau$$

so that

$$x(t) = e^{-3t} \int_0^t e^{3\tau} f(\tau)d\tau.$$

The function $f(t)$ is of course free to be assigned. In engineering and other applied subjects, $f(t)$ is made to take exotic forms; the discrete numbers corresponding to the output of laboratory measurements perhaps or even a time series with a stochastic (probabilistic) nature. However, here $f(t)$ must comply with our basic definition of a function and we must wait until Chapter 6 before we meet examples that involve discrete mathematics or statistically based functions. The ability to solve this kind of differential equation even with the definition of function met here has important practical consequences for the engineer and applied scientist. The function $f(t)$ is termed *input* and the term $x(t)$ *output*. To get from one to the other needs a *transfer function*. In the last example, the function $1/(s+3)$ written in terms of the transform variable s is this transfer function. This is the language of systems analysis, and such concepts also form the cornerstone of control engineering. They are also vital ingredients to branches of electrical engineering and the machine dynamics side of mechanical engineering. In mathematics, the procedure for writing the solution to a non-homogeneous differential equation (that is one with a non-zero right hand side) in terms of the solution of the corresponding homogeneous differential equation involves the development of the complementary function and particular solution. Complementary functions and particular solutions are standard concepts in solving second order ordinary differential equations, the subject of the next section.

3.3.1 Second Order Differential Equations

Let us now do a few examples to see how Laplace Transforms are used to solve *second* order ordinary differential equations. The technique is no different from solving first order ODEs, but finding the inverse Laplace Transform is often more challenging. Let us start by finding the solution to a homogeneous second order ODE that will be familiar to most of you who know about oscillations.

Example 3.17 *Use Laplace Transforms to solve the equation*

$$\frac{d^2x}{dt^2} + y = 0 \ with \ x(0) = 1, \ x'(0) = 0.$$

Solution Taking the Laplace Transform of this equation using the usual notation gives

$$s^2\bar{x}(s) - sx(0) - x'(0) + \bar{x}(s) = 0.$$

With $x(0) = 1$ and $x'(0) = 0$ we obtain

$$\bar{x}(s) = \frac{s}{s^2+1}.$$

This is a standard form which inverts to $x(t) = \cos(t)$. That this is the correct solution to this simple harmonic motion problem is easy to check.

Why not try changing the initial condition to $y(0) = 0$ and $y'(0) = 1$ which should lead to $y(t) = \sin(t)$? We are now ready to build on this result and solve the inhomogeneous problem that follows.

Example 3.18 *Find the solution to the differential equation*

$$\frac{d^2x}{dt^2} + x = t \ with \ x(0) = 1, \ x'(0) = 0.$$

Solution Apart from the trivial change of variable, we follow the last example and take Laplace Transforms to obtain

$$s^2\bar{x}(s) - sx(0) - x'(0) + \bar{x} = \mathcal{L}\{t\} = \frac{1}{s^2}.$$

With start conditions $x(0) = 1$ and $x'(0) = 0$ this gives

$$\bar{x}(s)(s^2+1) - s = \frac{1}{s^2}$$

$$so \ \bar{x}(s) = \frac{s}{s^2+1} + \frac{1}{s^2(s^2+1)}.$$

Taking the Inverse Laplace Transform thus gives:

$$x = \mathcal{L}^{-1}\left\{\frac{s}{s^2+1}\right\} + \mathcal{L}^{-1}\left\{\frac{1}{s^2(s^2+1)}\right\}$$
$$= \cos(t) + \int_0^t (t-\tau)\sin(\tau)d\tau$$

using the convolution theorem. Integrating by parts (omitting the details) gives

$$x = \cos(t) - \sin(t) + t.$$

The first two terms are the complementary function and the third the particular integral. The whole is easily checked to be the correct solution. It is up to the reader to decide whether this approach to solving this particular differential equation is any easier than the alternatives. The Laplace Transform method provides the solution of the differential equation with a *general* right hand side in a simple and straightforward manner.

However, instead of restricting attention to this particular second order differential equation, let us consider the more general equation

$$a\frac{d^2x}{dt^2} + b\frac{dx}{dt} + cx = f(t) \ (t \geq 0)$$

where a, b and c are constants. We will not solve this equation, but discuss it in the context of applications.

In engineering texts these constants are given names that have engineering significance. Although this text is primarily for a mathematical audience, it is nevertheless useful to run through these terms. In mechanics, a is the *mass*, b is the *damping constant* (diagrammatically represented by a dashpot), c is the *spring constant* (or *stiffness*) and x itself is the displacement of the mass. In electrical circuits, a is the *inductance*, b is the *resistance*, c is the reciprocal of the capacitance sometimes called the *reactance* and x (replaced by q) is the charge, the rate of change of which with respect to time is the more familiar electric current. Some of these names will be encountered later when we do applied examples. The right-hand side is called the *forcing* or *excitation*. In terms of systems engineering, $f(t)$ is the system input, and $x(t)$ is the system output. Since a, b and c are all constant the system described by the equation is termed *linear* and *time invariant*. It will seem very odd to a mathematician to describe a system governed by a time-dependent differential equation as "time invariant" but this is standard engineering terminology.

Taking the Laplace Transform of this general second order differential equation, assuming all the appropriate conditions hold of course, yields

$$a(s^2\bar{x}(s) - sx(0) - x'(0)) + b(s\bar{x}(s) - x(0)) + c\bar{x}(s) = \bar{f}(s).$$

It is normally not a problem to assume that $\bar{x}(s)$ and $\bar{f}(s)$ are of exponential order, but just occasionally when problems have a stochastic or numerical input, care needs to be taken. These kind of problems are met in Chapter 6. Making $\bar{x}(s)$ the subject of this equation gives

$$\bar{x}(s) = \frac{\bar{f}(s) + (as + b)x(0) + ax'(0)}{as^2 + bs + c}.$$

Hence, in theory, $x(t)$ can be found by taking Inverse Laplace Transform. The simplest case to consider is when $x(0)$ and $x'(0)$ are both zero. The output is then free from any embellishments that might be there because of special start conditions. In this special case,

$$\bar{x}(s) = \frac{1}{as^2 + bs + c}\bar{f}(s).$$

This equation is starkly in the form "response = transfer function × input" which makes it very clear why Laplace Transforms are highly regarded by engineers. The formula for $\bar{x}(s)$ can be inverted using the convolution theorem and examples of this can be found later in this chapter. First however let us solve a few simpler second order differential equations explicitly.

Example 3.19 *Use Laplace Transform techniques to find the solution to the second order differential equation*

$$\frac{d^2x}{dt^2} + 5\frac{dx}{dt} + 6x = 2e^{-t} \quad t \geq 0,$$

subject to the conditions $x = 1$ and $x' = 0$ at $t = 0$.

Solution Taking the Laplace Transform of this equation we obtain using the usual overbar notation,

$$s^2\bar{x}(s) - sx(0) - x'(0) + 5(s\bar{x}(s) - x(0)) + 6\bar{x}(s) = \mathcal{L}\{e^{-t}\} = \frac{2}{s+1}$$

where the standard Laplace Transform

$$\mathcal{L}\{e^{at}\} = \frac{1}{s-a}$$

for any constant a has been used. In future the so-called standard forms will not be quoted as they are listed in Appendix B. Inserting the initial conditions $x = 1$ and $x' = 0$ at $t = 0$ and rearranging the formula as an equation for $\bar{x}(s)$ gives

$$(s^2 + 5s + 6)\bar{x}(s) = \frac{2}{s+1} + s + 5.$$

Factorising gives

$$\bar{x}(s) = \frac{2}{(s+1)(s+2)(s+3)} + \frac{s+5}{(s+2)(s+3)}.$$

There are many ways of inverting this expression, the easiest being to use the partial fraction method of Chapter 2. Doing this but omitting the details gives:

$$\bar{x}(s) = \frac{1}{s+1} + \frac{1}{s+2} - \frac{1}{s+3}.$$

This inverts immediately to

$$x(t) = e^{-t} + e^{-2t} - e^{-3t} \quad (t \geq 0).$$

The first term is the particular integral (or particular solution) and the last two terms the complementary function. The whole solution is overdamped and therefore non-oscillatory: this is undeniably the easiest case to solve as it involves little algebra. However, it is also physically the least interesting as the solution dies away to zero very quickly. It does serve to demonstrate the power of the Laplace Transform technique to solve this kind of ordinary differential equation.

An obvious question to ask at this juncture is how is it known whether a particular inverse Laplace Transform be found? We know of course that it is obtainable in principle, but this is a practical question. In Chapter 7 we derive a general form for the inverse which helps to answer this question. Only a brief and informal answer can be given at this stage. As long as the function $\bar{f}(s)$ has a finite number of finite isolated singularities then inversion can go ahead. If $\bar{f}(s)$ does not tend to zero for large $|s|$ generalised functions are to be expected (see Chapter 2), and if $\bar{f}(s)$ has a square root or a more elaborate multi-valued nature then direct inversion is made more complicated, although there is no formal difficulty. In this case, error functions, Bessel functions and the like usually feature in the solution. Most of the time, solving second order linear differential equations is straightforward and involves no more than elementary transcendental functions (exponential and trigonometric functions).

The next problem is more interesting from a physical point of view.

Example 3.20 *Use Laplace Transforms to solve the following ordinary differential equation*

$$\frac{d^2x}{dt^2} + 6\frac{dx}{dt} + 9x = \sin(t) \quad (t \geq 0),$$

subject to $x(0) = 0$ *and* $x'(0) = 0$.

Solution With no right hand side, and with zero initial conditions, $x(t) = 0$ would result. However with the sinusoidal forcing, the solution turns out to be quite interesting. The formal way of tackling the problem is the same as for any second order differential equation with constant coefficients. Thus the mathematics follows that of the last example. Taking Laplace Transforms, the equation becomes

$$s^2\bar{x}(s) - sx(0) - x'(0) + 6s\bar{x}(s) - 6x(0) + 9\bar{x}(s) = \frac{1}{s^2+1}$$

this time not lingering over the standard form (for $\mathcal{L}\{\sin(t)\}$). With the boundary conditions inserted and a little tidying this becomes

$$\bar{x}(s) = \frac{1}{s^2 + 1} \cdot \frac{1}{(s+3)^2}.$$

Once again either partial fractions or convolution can be used. Convolution is our choice this time. Note that

$$\mathcal{L}\{\sin(t)\} = \frac{1}{s^2 + 1} \text{ and } \mathcal{L}\{te^{-3t}\} = \frac{1}{(s+3)^2}$$

so

$$\mathcal{L}^{-1}\left\{\frac{1}{(s^2+1)(s+3)^2}\right\} = \int_0^t \tau e^{-3\tau} \sin(t - \tau)d\tau.$$

This integral yields to integration by parts several times (or computer algebra, once). The result follows from application of the formula:

$$\int_0^t \tau e^{-3\tau} \sin(t - \tau)d\tau = -\frac{1}{10}\int_0^t e^{-3\tau}\cos(t-\tau)d\tau$$
$$+ \frac{3}{10}\int_0^t e^{-3\tau}\sin(t-\tau)d\tau + \frac{1}{10}te^{-3t}.$$

However we omit the details. The result is

$$x(t) = \mathcal{L}^{-1}\left\{\frac{1}{(s^2+1)(s+3)^2}\right\}$$
$$= \int_0^t \tau e^{-3\tau}\sin(t-\tau)d\tau$$
$$= \frac{e^{-3t}}{50}(5t+3) - \frac{3}{50}\cos(t) + \frac{2}{25}\sin(t).$$

The first term is the particular solution (called the *transient* response by engineers since it dies away for large times), and the final two terms the complementary function (rather misleadingly called the *steady state* response by engineers since it persists. Of course there is nothing steady about it!). After a "long time" has elapsed, the response is harmonic at the same frequency as the forcing frequency. The "long time" is in fact in practice quite short as is apparent from the graph of the output $x(t)$ which is displayed in Figure 3.3. The graph is indistinguishable from a sinusoid after about $t = 0.5$. However the amplitude and phase of the resulting oscillations are different. In fact, the combination

$$-\frac{3}{50}\cos(t) + \frac{2}{25}\sin(t) = \frac{1}{10}\sin(t - \epsilon)$$

puts the steady state into amplitude and phase form. $\frac{1}{10}$ is the amplitude and ϵ is the phase ($\cos(\epsilon) = \frac{4}{5}$, $\sin(\epsilon) = \frac{3}{5}$ for this solution). It is the fact that the

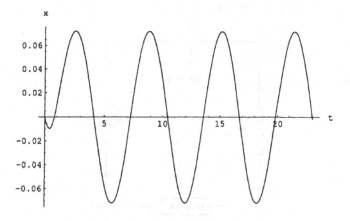

Figure 3.3: The graph of $x(t)$

response frequency is the same as the forcing frequency that is important in practical applications.

There is very little more to be said in terms of mathematics about the solution to second order differential equations with constant coefficients. The solutions are oscillatory, decaying, a mixture of the two, oscillatory and growing or simply growing exponentially. The forcing excites the response and if the response is at the same frequency as the natural frequency of the differential equation, resonance occurs. This leads to enhanced amplitudes at these frequencies. If there is no damping, then resonance leads to infinite amplitude response. Further details about the properties of the solution of second order differential equations with constant coefficients can be found in specialist books on differential equations and would be out of place here. What follows are examples where the power of the Laplace Transform technique is clearly demonstrated in terms of solving practical engineering problems. It is at the very applied end of applied mathematics. We return to a purer style in Chapter 4.

In the following example, a problem in electrical circuits is solved. As mentioned in the preamble to Example 3.19 the constants in a linear differential equation can be given significance in terms of the basic elements of an electrical circuit: *resistors, capacitors* and *inductors*. Resistors have resistance R measured in ohms, capacitors have capacitance C measured in farads, and inductors have inductance L measured in henrys. A current j flows through the circuit and the current is related to the charge q by

$$j = \frac{dq}{dt}.$$

The laws obeyed by a (passive) electrical circuit are:

1. *Ohm's law* whereby the voltage drop across a resistor is Rj.

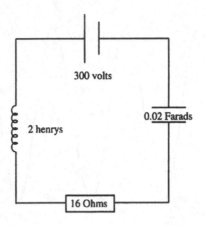

Figure 3.4: The simple circuit

2. The voltage drop across an inductor is

$$L\frac{dj}{dt}.$$

3. The voltage drop across a capacitor is

$$\frac{q}{C}.$$

Hence in terms of q the voltage drops are respectively

$$R\frac{dq}{dt}, \quad L\frac{d^2q}{dt^2} \quad \text{and} \quad \frac{q}{C}$$

which enables the circuit laws (Kirchhoff's Laws) to be expressed in terms of differential equations of second order with constant coefficients (L, R and $1/C$). The forcing function (input) on the right hand side is supplied by a voltage source, e.g. a battery. Here is a typical example.

Example 3.21 *Find the differential equation obeyed by the charge for the simple circuit shown in Figure 3.4, and solve it by the use of Laplace Transforms given $j = 0$, $q = 0$ at $t = 0$.*

Solution The current is j and the charge is q, so the voltage drop across the three devices are

$$2\frac{dj}{dt} = 2\frac{d^2q}{dt^2}, \quad 16j = 16\frac{dq}{dt}, \quad \text{and} \quad \frac{q}{0.002} = 50q.$$

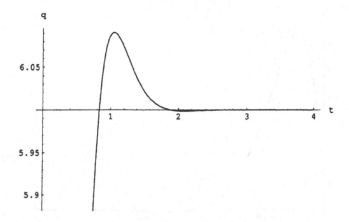

Figure 3.5: The solution $q(t)$; $q(0) = 0$ although true, is not shown due to the fine scale on the q axis

This must be equal to 300 (the voltage output of the battery), hence

$$2\frac{d^2q}{dt^2} + 16\frac{dq}{dt} + 50q = 300.$$

So, solving this by division by two and taking the Laplace Transform results in

$$s^2\bar{q}(s) - sq(0) - q'(0) + 8s\bar{q}(s) - 8q(0) + 25\bar{q}(s) = \frac{150}{s}.$$

Imposing the initial conditions gives the following equation for $\bar{q}(s)$, the Laplace Transform of $q(t)$:

$$\bar{q}(s) = \frac{150}{s(s^2 + 8s + 25)} = \frac{150}{s((s+4)^2 + 9)}.$$

This can be decomposed by using partial fractions or the convolution theorem. The former is easier. This gives:

$$\bar{q}(s) = \frac{6}{s} - \frac{6(s+4)}{(s+4)^2 + 9} - \frac{24}{(s+4)^2 + 9}.$$

The right hand side is now a standard form, so inversion gives:

$$q(t) = 6 - 6e^{-4t}\cos(3t) - 8e^{-4t}\sin(3t).$$

This solution is displayed in Figure 3.5. It can be seen that the oscillations are completely swamped by the exponential decay term. In fact, the current is the derivative of this which is:-

$$j = 50e^{-4t}\sin(3t)$$

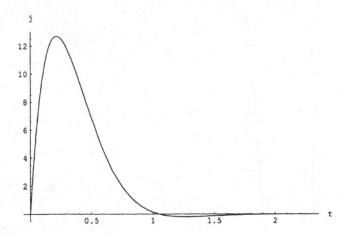

Figure 3.6: The variation of current with time

and this is shown in Figure 3.6 as decaying to zero very quickly. This is obviously not typical, as demonstrated by the next example. Here, we have a sinusoidal voltage source which might be thought of as mimicking the production of an alternating current.

Example 3.22 *Solve the same problem as in the previous example except that the battery is replaced by the oscillatory voltage source* $100\sin(3t)$.

Solution The differential equation is derived as before, except for the different right hand side. The equation is

$$2\frac{d^2q}{dt^2} + 16\frac{dq}{dt} + 50q = 100\sin(3t).$$

Taking the Laplace Transform of this using the zero initial conditions $q(0) = 0, q'(0) = 0$ proceeds as before, and the equation for the Laplace Transform of $q(t)$ $(\bar{q}(s))$ is

$$\bar{q}(s) = \frac{150}{(s^2+9)((s+4)^2+9)}.$$

Note the appearance of the Laplace Transform of $100\sin(3t)$ here. The choice is to either use partial fractions or convolution to invert, this time we use convolution, and this operation by its very nature recreates $100\sin(3t)$ under an integral sign "convoluted" with the complementary function of the differential equation. Recognising the two standard forms:-

$$\mathcal{L}^{-1}\left\{\frac{1}{s^2+9}\right\} = \frac{1}{3}\sin(3t) \text{ and } \mathcal{L}^{-1}\left\{\frac{1}{(s+4)^2+9}\right\} = \frac{1}{3}e^{-4t}\sin(3t),$$

gives immediately

$$q(t) = \frac{50}{3}\int_0^t e^{-4(t-\tau)}\sin\left(3(t-\tau)\right)\sin(3\tau)d\tau.$$

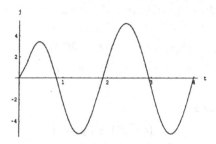

Figure 3.7: The variation of current with time

Using integration by parts but omitting the details gives

$$J = \frac{dq(t)}{dt} = \frac{75}{52}(2\cos(3t) + 3\sin(3t)) - \frac{25}{52}e^{-4t}(17\sin(3t) + 6\cos(3t)).$$

What this solution tells the electrical engineer is that the response quickly becomes sinusoidal at the same frequency as the forcing function but with smaller amplitude and different phase. This is backed up by glancing at Figure 3.7 which displays this solution. The behaviour of this solution is very similar to that of the mechanical engineering example, Example 3.25, that we will soon meet and demonstrates beautifully the merits of a mathematical treatment of the basic equations of engineering. A mathematical treatment enables analogies to be drawn between seemingly disparate branches of engineering.

3.3.2 Simultaneous Differential Equations

In the same way that Laplace Transforms convert a single differential equation into a single algebraic equation, so they can also convert a pair of differential equations into simultaneous algebraic equations. The differential equations we solve are all linear, so a pair of linear differential equations will convert into a pair of simultaneous linear algebraic equations familiar from school. Of course, these equations will contain s, the transform variable as a parameter. These expressions in s can get quite complicated. This is particularly so if the forcing functions on the right-hand side lead to algebraically involved functions of s. Comments on the ability or otherwise of inverting these expressions remain the same as for a single differential equation. They are more complicated, but still routine. Let us start with a straightforward example.

Example 3.23 *Solve the simultaneous differential equations*

$$\frac{dx}{dt} = 2x - 3y, \frac{dy}{dt} = y - 2x,$$

where $x(0) = 8$ and $y(0) = 3$.

Solution Taking Laplace Transforms and inserting the boundary conditions straight away gives:-

$$s\bar{x}(s) - 8 = 2\bar{x}(s) - 3\bar{y}(s)$$
$$s\bar{y}(s) - 3 = \bar{y}(s) - 2\bar{x}(s).$$

Whence, rearranging we solve:-

$$(s - 2)\bar{x} + 3\bar{y} = 8$$
$$\text{and} \quad 2\bar{x} + (s - 1)\bar{y} = 3$$

by the usual means. Using Cramer's rule or eliminating by hand gives the solutions

$$\bar{x}(s) = \frac{8s - 17}{s^2 - 3s - 4}, \ \bar{y}(s) = \frac{3s - 22}{s^2 - 3s - 4}.$$

To invert these we factorise and decompose into partial fractions to give

$$\bar{x}(s) = \frac{5}{s+1} + \frac{3}{s-4}$$
$$\bar{y}(s) = \frac{5}{s+1} - \frac{2}{s-4}.$$

These invert easily and we obtain the solution

$$x(t) = 5e^{-t} + 3e^{4t}$$
$$y(t) = 5e^{-t} - 2e^{4t}.$$

Even if one or both of these equations were second order, the solution method by Laplace Transforms remains the same. The following example hints at how involved the algebra can get, even in the most innocent looking pair of equations!

Example 3.24 *Solve the simultaneous differential equations*

$$\frac{d^2x}{dt^2} + \frac{dy}{dt} + 3x = 15e^{-t}$$

$$\frac{d^2y}{dt^2} - 4\frac{dx}{dt} + 3y = 15\sin(2t)$$

where $x = 35, x' = -48, y = 27$ *and* $y' = -55$ *at time* $t = 0$.

Solution Taking Laplace Transforms of both equations as before, retaining the standard notation for the transformed variable gives

$$s^2\bar{x} - 35s + 48 + s\bar{y} - 27 + 3\bar{x} = \frac{15}{s+1}$$

$$\text{and} \ s^2\bar{y} - 27s + 55 - 4(s\bar{x} - 35) + 3\bar{y} = \frac{30}{s^2 + 4}.$$

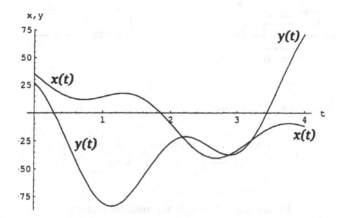

Figure 3.8: The solutions $x(t)$ and $y(t)$

Solving for $\bar{x}$ and $\bar{y}$ is indeed messy but routine. We do one step to group the terms together as follows:

$$(s^2 + 3)\bar{x} + s\bar{y} - 35s - 21 + \frac{15}{s+1}$$

$$-4s\bar{x} + (s^2 + 3)\bar{y} = 27s - 195 + \frac{30}{s^2 + 4}.$$

This time, this author has used a computer algebra package to solve these two equations. A partial fraction routine has also been used. The result is

$$\bar{x}(s) = \frac{30s}{s^2 + 1} - \frac{45}{s^2 + 9} + \frac{3}{s+1} + \frac{2s}{s^2 + 4}$$

$$\text{and } \bar{y}(s) = \frac{30s}{s^2 + 9} - \frac{60}{s^2 + 1} - \frac{3}{s+1} + \frac{2}{s^2 + 4}.$$

Inverting using standard forms gives

$$x(t) = 30\cos(t) - 15\sin(3t) + 3e^{-t} + 2\cos(2t)$$

$$y(t) = 30\cos(3t) - 60\sin(t) - 3e^{-t} + \sin(2t).$$

The last two terms on the right-hand side of the expression for both x and y resemble the forcing terms whilst the first two are in a sense the "complementary function" for the system. The motion is quite a complex one and is displayed as Figure 3.8

Figure 3.9: The forces due to **a** a damper and **b** a spring

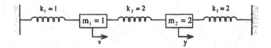

Figure 3.10: A simple mechanical system

Having looked at the application of Laplace Transforms to electrical circuits, now let us apply them to mechanical systems. Again it is emphasised that there is no new mathematics here; it is however new *applied* mathematics.

In mechanical systems, we use Newton's second law to determine the motion of a mass which is subject to a number of forces. The kind of system best suited to Laplace Transforms are the mass-spring-damper systems. Newton's second law is of the form

$$F = m\frac{d^2x}{dt^2}$$

where F is the force, m is the mass and x the displacement. The components of the system that also act on the mass m are a *spring* and a *damper*. Both of these give rise to changes in displacement according to the following rules (see Figure 3.9). A damper produces a force proportional to the net speed of the mass but always opposes the motion, i.e. $c(\dot{y} - \dot{x})$ where c is a constant and the dot denotes differentiation with respect to t. A spring produces a force which is proportional to displacement. Here, springs will be well behaved and assumed to obey Hooke's Law. This force is $k(y - x)$ where k is a constant sometimes called the *stiffness* by mechanical engineers. To put flesh on these bones, let us solve a typical mass spring damping problem. Choosing to consider two masses gives us the opportunity to look at an application of simultaneous differential equations.

Example 3.25 *Figure 3.10 displays a mechanical system. Find the equations of motion, and solve them given that the system is initially at rest with $x = 1$ and $y = 2$.*

Solution Applying Newton's Second Law of Motion successively to each mass using Hooke's Law (there are no dampers) gives:-

$$m_1\ddot{x} = k_2(y - x) - k_1x$$
$$m_2\ddot{y} = -k_3x - k_2(y - x).$$

With the values for the constants m_1, m_2, k_1, k_2 and k_3 given in Figure 3.10, the following differential equations are obtained:-

$$\ddot{x} + 3x - 2y = 0$$
$$2\ddot{y} + 4y - 2x = 0.$$

The mechanics (thankfully for most) is now over, and we take the Laplace Transform of both equations to give:-

$$(s^2 + 3)\bar{x} - 2\bar{y} = sx(0) + \dot{x}(0)$$
$$-\bar{x} + (s^2 + 2)\bar{y} = sy(0) + \dot{y}(0).$$

The right hand side involves the initial conditions which are: $x(0) = 1$, $y(0) = 2$, $\dot{x}(0) = 0$ and $\dot{y}(0) = 0$. Solving these equations (by computer algebra or by hand) gives, for $\bar{y}$,

$$\bar{y}(s) = \frac{2s^3 + 5s}{(s^2 + 4)(s^2 + 1)} = \frac{s}{s^2 + 1} + \frac{s}{s^2 + 4}$$

and inverting gives

$$y(t) = \cos(t) + \cos(2t).$$

Rather than finding $\bar{x}$ and inverting, it is easier to substitute for y and its second derivative $\ddot{y}$ in the equation

$$x = \ddot{y} + 2y.$$

This finds x directly as

$$x(t) = \cos(t) - 2\cos(2t).$$

The solution is displayed as Figure 3.11. It is possible and mechanically desirable to form the combinations

$$\frac{1}{3}(y - x) = \cos(2t) \text{ and } \frac{1}{3}(x + 2y) = \cos(t)$$

as these isolate the two frequencies and help in the understanding of the subsequent motion which at first sight can seem quite complex. This introduces the concept of *normal modes* which are outside the scope of this text, but very important to mechanical engineers as well as anyone else interested in the behaviour of oscillating systems.

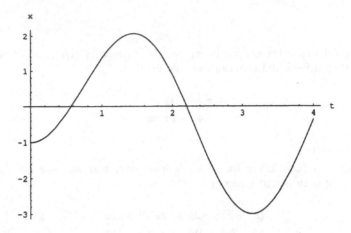

Figure 3.11: A simple mechanical system solved

3.4   Using Step and Impulse Functions

In Section 2.3 some properties of Heaviside's Unit Step Function were explored. In this section we extend this exploration to problems that involve differential equations. As a reminder the step function $H(t)$ is defined as

$$H(t) = \begin{cases} 0 & t < 0 \\ 1 & t \geq 0 \end{cases}$$

and a function $f(t)$ which is switched on at time t_0 is represented simply as $f(t - t_0)H(t - t_0)$. The second shift theorem, Theorem 2.5, implies that the Laplace Transform of this is given by

$$\mathcal{L}\{f(t - t_0)H(t - t_0)\} = e^{-st_0}\bar{f}(s).$$

Another useful result derived in Chapter 2 is

$$\mathcal{L}\{\delta(t - a)f(t)\} = e^{-as}f(a)$$

where $\delta(t - a)$ is the Dirac -δ or impulse function centred at $t = a$. The properties of the δ function as required in this text are outlined in Section 2.6. Let us use these properties to solve an engineering problem. This next example is quite extensive; more of a case study. It involves concepts usually found in mechanics texts, although it is certainly possible to solve the differential equation that is obtained abstractly and without recourse to mechanics: much would be missed in terms of realising the practical applications of Laplace Transforms. Nevertheless this example can certainly be omitted from a first reading, and discarded entirely by those with no interest in applications to engineering.

Example 3.26 *The equation governing the bending of beams is of the form*

$$k\frac{d^4y}{dx^4} = -W(x)$$

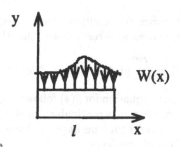

Figure 3.12: The beam and its load $W(x)$

where k is a constant called the flexural rigidity (the product of Young's Modulus and length in fact, but this is not important here), and $W(x)$ is the transverse force per unit length along the beam. The layout is indicated in Figure 3.12. Use Laplace Transforms (in x) to solve this problem and discuss the case of a point load.

Solution There are several aspects to this problem that need a comment here. The mathematics comes down to solving an ordinary differential equation which is fourth order but easy enough to solve. In fact, only the fourth derivative of $y(x)$ is present, so in normal circumstances one might expect direct integration (four times) to be possible. That it is not is due principally to the form $W(x)$ usually takes. There is also that the beam is of finite length l. In order to use Laplace Transforms the domain is extended so that $x \in [0, \infty)$ and the Heaviside Step Function is utilised. To progress in a step by step fashion let us consider the cantilever problem first where the beam is held at one end. Even here there are conditions imposed at the free end. However, we can take Laplace Transforms in the usual way to eliminate the x derivatives. We define the Laplace Transform in x as

$$\bar{y}(s) = \int_0^\infty e^{-xs} y(x) dx$$

where remember we have extended the domain to ∞. In transformed coordinates the equation for the beam becomes:-

$$k(s^4 \bar{y}(s) - s^3 y(0) - s^2 y'(0) - s y''(0) - y'''(0)) = -\overline{W}(s).$$

Thus,

$$\bar{y}(s) = \frac{-\overline{W}(s)}{ks^4} + \frac{y(0)}{s} + \frac{y'(0)}{s^2} + \frac{y''(0)}{s^3} + \frac{y'''(0)}{s^4}$$

and the solution can be found by inversion. It is at this point that the engineer would be happy, but the mathematician should be pausing for thought! The beam may be long, but it is not infinite. This being the case, is it legitimate to define the Laplace Transform in x as has been done here? What needs to be done is some tidying up using Heaviside's Step Function. If we replace $y(x)$ by the combination $y(x)[1 - H(x - l)]$ then this latter function will certainly fulfil the necessary and sufficient conditions for the existence of the Laplace Transform provided $y(x)$ is piecewise continuous. One therefore interprets $\bar{y}(s)$ as

$$\bar{y}(s) = \int_0^\infty y(x)[1 - H(x - l)]e^{-xs}dx$$

and inversion using the above equation for $\bar{y}(s)$ follows once the forcing is known. In general, the convolution theorem is particularly useful here as $W(x)$ may take the form of data (from a strain gauge perhaps) or have a stochastic character. Using the convolution theorem, we have

$$\mathcal{L}^{-1}\left\{\frac{\overline{W}(s)}{s^4}\right\} = \frac{1}{6}\int_0^x (x - \xi)^3 W(\xi)d\xi.$$

The solution to the problem is therefore

$$y(x)[1 - H(x - l)] = -\frac{1}{6k}\int_0^x (x - \xi)^3 W(\xi)d\xi$$
$$+ y(0) + xy'(0) + \frac{1}{2}x^2y''(0) + \frac{1}{6}x^3y'''(0).$$

If the beam is freely supported at both ends, this is mechanics code for the following four boundary conditions

$$y = 0 \text{ at } x = 0, \ l \text{ (no displacement at the ends)}$$

and

$$y'' = 0 \text{ at } x = 0, \ l \text{ (no force at the ends)}.$$

This enables the four constants of integration to be found. Straightforwardly, $y(0) = 0$ and $y''(0) = 0$ so

$$y(x)[1 - H(x - l)] = -\frac{1}{6k}\int_0^x (x - \xi)^3 W(\xi)d\xi + xy'(0) + \frac{1}{6}x^3y'''(0).$$

The application of boundary conditions at $x = l$ is less easy. One method would be to differentiate the above expression with respect to x twice, but it is unclear how to apply this to the product on the left, particularly at $x = l$. The following procedure is recommended. Put $u(x) = y''(x)$ and the original differential equation, now in terms of $u(x)$ becomes

$$k\frac{d^2u}{dx^2} = -W(x)$$

with solution (obtained by using Laplace Transforms as before) given by

$$u(x)[1 - H(x - l)] = -\frac{1}{k}\int_0^x (x - \xi)W(\xi)d\xi + u(0) + xu'(0).$$

Hence the following expression for $y''(x)$ has been derived

$$y''(x)[1 - H(x - l)] = -\frac{1}{6}\int_0^x (x - \xi)W(\xi)d\xi + y''(0) + xy'''(0).$$

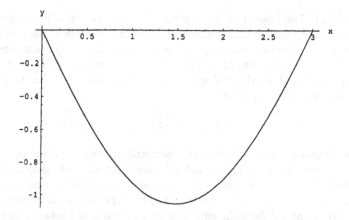

Figure 3.13: The displacement of the beam $y(x)$ with $W = 1$ and $k = 1$. The length l equals 3.

This is in fact the result that would have been obtained by differentiating the expression for $y(x)$ twice ignoring derivatives of $[1 - H(x - l)]$. Applying the boundary conditions at $x = l$ ($y(l) = y''(l) = 0$) now gives the results

$$y'''(0) = \frac{1}{kl} \int_0^l (l - \xi)W(\xi)d\xi$$

and

$$y'(0) = \frac{1}{6kl} \int_0^l (l - \xi)^3 W(\xi)d\xi - \frac{l}{6k} \int_0^l (l - \xi)W(\xi)d\xi.$$

This provides the general solution to the problem in terms of integrals

$$y(x)[1 - H(x - l)] = -\frac{1}{6k} \int_0^x (x - \xi)^3 W(\xi)d\xi$$

$$+ \frac{1}{6kl} \int_0^l x(l - \xi)(\xi^2 - 2l\xi + x^2)W(\xi)d\xi.$$

It is now possible to insert any loading function into this expression and calculate the displacement caused. In particular, let us choose the values $l = 3$, $k = 1$ and consider a uniform loading of unit magnitude $W = \text{constant} = 1$. The integrals are easily calculated in this simple case and the resulting displacement is

$$y(x) = -\frac{x^4}{24} + \frac{x^3}{4} - \frac{9x}{8}$$

which is shown in Figure 3.13.

We now consider a more realistic loading, that of a point load located at $x = l/3$ (one third of the way along the bar). It is now the point at which the laws of mechanics need to be applied in order to translate this into a specific form for $W(x)$. This however is not a mechanics text, therefore it is quite likely that you are not familiar with enough of these laws to follow the derivation. For those with mechanics knowledge, we assume that the weight of the beam is concentrated at its mid-point ($x = l/2$) and that the beam itself is static so that there is no turning about the free end at $x = l$. If the point load has magnitude P, then the expression for $W(x)$ is

$$W(x) = \frac{W}{l} H(x) + P\delta\left(x - \frac{1}{3}l\right) - \left(\frac{1}{2}W + \frac{2}{3}P\right)\delta(x).$$

From a mathematical point of view, the interesting point here is the presence of the Dirac-δ function on the right hand side which means that integrals have to be handled with some care. For this reason, and in order to present a different way of solving the problem but still using Laplace Transforms we go back to the fourth order ordinary differential equation for $y(x)$ and take Laplace Transforms. The Laplace Transform of the right hand side $(W(x))$ is

$$\overline{W}(s) = \frac{W}{ls} + Pe^{-\frac{1}{3}sl} - (\frac{1}{2}W + \frac{2}{3}P).$$

The boundary conditions are $y = 0$ at $x = 0,\ l$ (no displacement at the ends) and $y''(0) = 0$ at $x = 0,\ l$ (no forces at the ends). This gives

$$\bar{y}(s) = \frac{1}{k}\left[\frac{W}{ls^5} + \frac{P}{s^4}e^{-\frac{1}{3}ls} - \frac{1}{s^4}\left(\frac{1}{2}W + \frac{2}{3}P\right)\right] + \frac{y'(0)}{s^2} + \frac{y'''(0)}{s^4}.$$

This can be inverted easily using standard forms, together with the second shift theorem for the exponential term to give:-

$$y(x)[1 - H(x - l)] = -\frac{1}{k}\left[\frac{W}{24l}x^4 + \frac{1}{6}P\left(x - \frac{1}{3}l\right)^3 - \frac{1}{6}\left(\frac{1}{2}W + \frac{2}{3}P\right)x^3\right]$$

$$+ y'(0)x + \frac{1}{6}y'''(0)x^3.$$

Differentiating twice gives

$$y''(x) = -\frac{1}{k}\left[\frac{1}{2l}Wx^2 + P\left(x - \frac{1}{3}l\right) - \left(\frac{1}{2}W + \frac{2}{3}P\right)x\right] + y'''(0)x, \quad 0 \le x \le l.$$

This is zero at $x = l$, whence $y'''(0) = 0$. The boundary condition $y(l) = 0$ is messier to apply as it is unnatural for Laplace Transforms. It gives

$$y'(0) = -\frac{l^2}{k}\left(\frac{1}{24}W + \frac{5}{81}P\right)$$

so the solution valid for $0 \le x \le l$ is

$$y(x) = -\frac{W}{k}\left(\frac{1}{24l}x^4 - \frac{1}{12}x^3 + \frac{1}{24}l^2x\right) - \frac{P}{k}\left(\frac{5}{81}l^2x - \frac{1}{9}x^3\right)$$

$$- \frac{P}{6k}\left(x - \frac{1}{3}l\right)^3 H\left(x - \frac{1}{3}l\right).$$

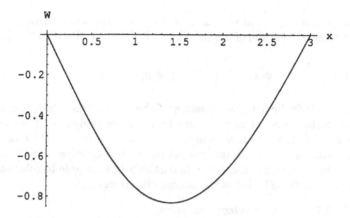

Figure 3.14: The displacement of the beam $y(x)$ with $W = 1, k = 1$ and $P = 1$. The length l equals 3

This solution is illustrated in Figure 3.14

Other applications to problems that give rise to partial differential equations will have to wait until Chapter 5.

3.5   Integral Equations

An integral equation is, as the name implies, an equation in which the unknown occurs in the form of an integral. The Laplace Transform proves very useful in solving many types of integral equation, especially when the integral takes the form of a convolution. Here are some typical integral equations:

$$\int_a^b K(x,y)\phi(y)dy = f(x)$$

where K and f are known. A more general equation is:

$$A(x)\phi(x) - \lambda \int_a^b K(x,y)\phi(y)dy = f(x)$$

and an even more general equation would be:

$$\int_a^b K(x, y, \phi(y))dy = f(x).$$

In these equations, K is called the kernel of the integral equation. The general theory of how to solve integral equations is outside the scope of this text, and we shall content ourselves with solving a few special types particularly suited to solution using Laplace Transforms. The last very general integral equation is non-linear and is in general very difficult to solve. The second type is called a Fredholm integral equation of the third kind. If $A(x) = 1$, this equation becomes

$$\phi(x) - \lambda \int_a^b K(x,y)\phi(y)dy = f(x)$$

which is the Fredholm integral equation of the second kind. In integral equations, x is the independent variable, so a and b can depend on it. The Fredholm integral equation covers the case where a and b are constant, in the cases where a or b (or both) depend on x the integral equation is called a Volterra integral equation. One particular case, $b = x$, is particularly amenable to solution using Laplace Transforms. The following example illustrates this.

Example 3.27 *Solve the integral equation*

$$\phi(x) - \lambda \int_0^x e^{x-y}\phi(y)dy = f(x)$$

where $f(x)$ is a general function of x.

Solution The integral is in the form of a convolution; it is in fact $\lambda e^x * \phi(x)$ where * denotes the convolution operation. The integral can thus be written

$$\phi(x) - \lambda e^x * \phi(x) = f(x).$$

Taking the Laplace Transform of this equation and utilising the convolution theorem gives

$$\bar{\phi} - \lambda \frac{\bar{\phi}}{s-1} = \bar{f}$$

where $\bar{\phi} = \mathcal{L}\phi$ and $\bar{f} = \mathcal{L}f$. Solving for $\bar{\phi}$ gives

$$\bar{\phi}\left(\frac{s-1-\lambda}{s-1}\right) = \bar{f}$$

$$\bar{\phi} = \frac{s-1}{s-(1+\lambda)}\bar{f} = \bar{f} + \frac{\lambda\bar{f}}{s-(1+\lambda)}.$$

So inverting gives

$$\phi = f + \lambda f * e^{(1+\lambda)x}$$

and

$$\phi(x) = f(x) + \lambda \int_0^x f(y)e^{(1+\lambda)(x-y)}dy.$$

This is the solution of the integral equation.

The solution of integral equations of these types usually involves advanced methods including complex variable methods. These can only be understood after the methods of Chapter 7 have been introduced and are the subject of more advanced texts (e.g. Hochstadt (1989)).

3.6 Exercises

1. Given suitably well behaved functions f, g and h establish the following properties of the convolution $f * g$ where

$$f * g = \int_0^t f(\tau)g(t - \tau)d\tau.$$

(a) $f * g = g * f$, (b) $f * (g * h) = (f * g) * h$,
(c) determine f^{-1} such that $f * f^{-1} = 1$, stating any extra properties f must possess in order for the inverse f^{-1} to exist.

2. Use the convolution theorem to establish

$$\mathcal{L}^{-1}\left\{\frac{\bar{f}(s)}{s}\right\} = \int_0^t f(\tau)d\tau.$$

3. Find the following convolutions

(a) $t * \cos(t)$, (b) $t * t$, (c) $\sin(t) * \sin(t)$, (d) $e^t * t$, (e) $e^t * \cos(t)$.

4. (a) Show that

$$\lim_{x \to 0} \left(\frac{\text{erf}(x)}{x}\right) = \frac{2}{\sqrt{\pi}}.$$

(b) Show that

$$\mathcal{L}\{t^{-1/2}\text{erf}\sqrt{t}\} = \frac{2}{\sqrt{\pi s}} \tan^{-1}\left(\frac{1}{\sqrt{s}}\right).$$

5. Solve the following differential equations by using Laplace Transforms:
(a)

$$\frac{dx}{dt} + 3x = e^{2t}, \quad x(0) = 1,$$

(b)

$$\frac{dx}{dt} + 3x = \sin(t), \quad x(\pi) = 1,$$

(c)

$$\frac{d^2x}{dt^2} + 4\frac{dx}{dt} + 5x = 8\sin(t), \quad x(0) = x'(0) = 0,$$

(d)

$$\frac{d^2x}{dt^2} - 3\frac{dx}{dt} - 2x = 6, \quad x(0) = x'(0) = 1,$$

(e)

$$\frac{d^2y}{dt^2} + y = 3\sin(2t), \quad y(0) = 3, \quad y'(0) = 1.$$

6. Solve the following pairs of simultaneous differential equations by using Laplace Transforms:

(a)

$$\frac{dx}{dt} - 2x - \frac{dy}{dt} - y = 6e^{3t}$$

$$2\frac{dx}{dt} - 3x + \frac{dy}{dt} - 3y = 6e^{3t}$$

subject to $x(0) = 3$, and $y(0) = 0$.

(b)

$$4\frac{dx}{dt} + 6x + y = 2\sin(2t)$$

$$\frac{d^2x}{dt^2} + x - \frac{dy}{dt} = 3e^{-2t}$$

subject to $x(0) = 2, x'(0) = -2$, (eliminate y).

(c)

$$\frac{d^2x}{dt^2} - x + 5\frac{dy}{dt} = t$$

$$\frac{d^2y}{dt^2} - 4y - 2\frac{dx}{dt} = -2$$

subject to $x(0) = 0, x'(0) = 0, y(0) = 1$ and $y'(0) = 0$.

7. Demonstrate the phenomenon of *resonance* by solving the equation:

$$\frac{d^2x}{dt^2} + k^2x = A\sin(t)$$

subject to $x(0) = x_0, \quad x'(0) = v_0$ then showing that the solution is unbounded as $t \to \infty$ for particular values of k.

8. Assuming that the air resistance is proportional to speed, the motion of a particle in air is governed by the equation:

$$\frac{w}{g}\frac{d^2x}{dt^2} + b\frac{dx}{dt} = w.$$

If at $t = 0, x = 0$ and

$$v = \frac{dx}{dt} = v_0$$

show that the solution is

$$x = \frac{gt}{a} + \frac{(av_0 - g)(1 - e^{-at})}{a^2}$$

where $a = bg/w$. Deduce the terminal speed of the particle.

9. Determine the algebraic system of equations obeyed by the transformed electrical variables $\bar{j}_1, \bar{j}_2, \bar{j}_3$ and $\bar{q}_3$ given the electrical circuit equations

$$j_1 = j_2 + j_3$$

$$R_1 j_1 + L_2 \frac{dj_2}{dt} = E\sin(\omega t)$$

$$R_1 j_1 + R_3 j_3 + \frac{1}{C} q_3 = E\sin(\omega t)$$

$$j_3 = \frac{dq_3}{dt}$$

where, as usual, the overbar denotes the Laplace Transform.

10. Solve the loaded beam problem expressed by the equation

$$k\frac{d^4 y}{dx^4} = \frac{w_0}{c}[c - x + (x - c)H(x - c)] \quad \text{for } 0 < x < 2c$$

subject to the boundary conditions $y(0) = 0$, $y'(0) = 0$, $y''(2c) = 0$ and $y'''(2c) = 0$. $H(x)$ is the Heaviside Unit Step Function. Find the bending moment ky'' at the point $x = \frac{1}{2}c$.

11. Solve the integral equation

$$\phi(t) = t^2 + \int_0^t \phi(y)\sin(t - y)dy.$$

$$4$$

Fourier Series

4.1 Introduction

Before getting to Fourier series proper, we need to discuss the context. To understand why Fourier series are so useful, one would need to define an inner product space and show that trigonometric functions are an example of one. It is the properties of the inner product space, coupled with the analytically familiar properties of the sine and cosine functions that give Fourier series their usefulness and power. Some familiarity with set theory, vector and linear spaces would be useful. These are topics in the first stages of most mathematical degrees, but if they are new, the text by Whitelaw (1983) will prove useful.

The basic assumption behind Fourier series is that any given function can be expressed in terms of a series of sine and cosine functions, and that once found the series is unique. Stated coldly with no preliminaries this sounds preposterous, but to those familiar with the theory of linear spaces it is not. All that is required is that the sine and cosine functions are a *basis* for the linear space of functions to which the given function belongs. Some details are given in Appendix C. Those who have a background knowledge of linear algebra sufficient to absorb this appendix should be able to understand the following two theorems which are essential to Fourier series. They are given without proof and may be ignored by those willing to accept the results that depend on them. The first result is Bessel's inequality. It is conveniently stated as a theorem.

Theorem 4.1 (Bessel's Inequality) *If*

$$\{e_1, e_2, \ldots, e_n, \ldots\}$$

79

is an orthonormal basis for the linear space V, then for each $\mathbf{a} \in V$ the series

$$\sum_{r=1}^{\infty} |\langle \mathbf{a}, \mathbf{e}_n \rangle|^2$$

converges. In addition, the inequality

$$\sum_{r=1}^{\infty} |\langle \mathbf{a}, \mathbf{e}_n \rangle|^2 \leq ||\mathbf{a}||^2$$

holds.

An important consequence of Bessel's inequality is the Riemann–Lebesgue lemma. This is also stated as a theorem:-

Theorem 4.2 (Riemann–Lebesgue) *Let $\{\mathbf{e}_1, \mathbf{e}_2, \ldots\}$ be an orthonormal basis of infinite dimension for the inner product space V. Then, for any $\mathbf{a} \in V$*

$$\lim_{n \to \infty} \langle \mathbf{a}, \mathbf{e}_n \rangle = 0.$$

This theorem in fact follows directly from Bessel's inequality as the nth term of the series on the right of Bessel's inequality must tend to zero as n tends to ∞.

Although some familiarity with analysis is certainly a prerequisite here, there is merit in emphasising the two concepts of pointwise convergence and uniform convergence. It will be out of place to go into proofs, but the difference is particularly important to the study of Fourier series as we shall see later. Here are the two definitions.

Definition 4.3 (Pointwise Convergence) *Let*

$$\{f_0, f_1, \ldots, f_m, \ldots\}$$

be a sequence of functions defined on the closed interval $[a, b]$. We say that the sequence $\{f_0, f_1, \ldots, f_m, \ldots\}$ converges pointwise to f on $[a, b]$ if for each $x \in [a, b]$ and $\epsilon > 0$ there exists a natural number $N(\epsilon, x)$ such that

$$|f_m(x) - f(x)| < \epsilon$$

for all $m \geq N(\epsilon, x)$.

Definition 4.4 (Uniform Convergence) *Let*

$$\{f_0, f_1, \ldots, f_m, \ldots\}$$

be a sequence of functions defined on the closed interval $[a, b]$. We say that the sequence $\{f_0, f_1, \ldots, f_m, \ldots\}$ converges uniformly to f on $[a, b]$ if for each $\epsilon > 0$ there exists a natural number $N(\epsilon)$ such that

$$|f_m(x) - f(x)| < \epsilon$$

for all $m \geq N(\epsilon)$ and for all $x \in [a, b]$.

It is the difference and not the similarity of these two definitions that is important. All uniformly convergent sequences are pointwise convergent, but not vice versa. This is because N in the definition of pointwise convergence depends on x; in the definition uniform convergence it does not which makes uniform convergence a global rather than a local property. The N in the definition of uniform convergence will do for any x in $[a, b]$.

Armed with these definitions and assuming a familiarity with linear spaces, we will eventually go ahead and find the Fourier series for a few well known functions. We need a few more preliminaries before we can do this.

4.2   Definition of a Fourier Series

As we have said, Fourier series consist of a series of sine and cosine functions. We have also emphasised that the theory of linear spaces can be used to show that it possible to represent any periodic function to any desired degree of accuracy provided the function is periodic and piecewise continuous (see Appendix C for some details). To start, it is easiest to focus on functions that are defined in the closed interval $[-\pi, \pi]$. These functions will be piecewise continuous and they will possess one sided limits at $-\pi$ and π. So, using mathematical notation, we have $f : [-\pi, \pi] \to \mathbf{C}$. The restriction to this interval will be lifted later, but periodicity will always be essential.

It also turns out that the points at which f is discontinuous need not be points at which f is defined uniquely. As an example of what is meant, Figure 4.1 shows three possible values of the function

$$f_a = \left\{ \begin{array}{ll} 0 & t < 1 \\ 1 & t > 1 \end{array} \right.$$

at $t = 1$. These are $f_a(1) = 0, f_a(1) = 1$ and $f_a(1) = 1/2$, and, although we do need to be consistent in order to satisfy the need for $f_a(t)$ to be well defined, in theory it does not matter exactly where $f_a(1)$ is. However, Figure 4.1c is the right choice for Fourier series; the following theorem due to Dirichlet tells us why.

Theorem 4.5 *If f is a member of the space of piecewise continuous functions which are 2π periodic on the closed interval $[-\pi, \pi]$ and which has both left and right derivatives at each $x \in [-\pi, \pi]$, then for each $x \in [-\pi, \pi]$ the Fourier series of f converges to the value*

$$\frac{f(x_-) + f(x_+)}{2}.$$

At both end points, $x = \pm\pi$, the series converges to

$$\frac{f(\pi_-) + f((-\pi)_+)}{2}.$$

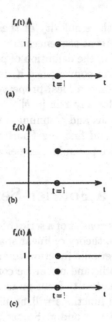

Figure 4.1: (a) $f_a(1) = 0$,(b) $f_a(1) = 1$,(c) $f_a(1) = 1/2$

The proof of this is beyond the scope of this text, but some comments are usefully made. If x is a point at which the function f is continuous, then

$$\frac{f(x_-) + f(x_+)}{2} = f(x)$$

and the theorem is certainly eminently plausible as any right hand side other than $f(x)$ for this mean of left and right sided limits would be preposterous. It is still however difficult to prove rigorously. At other points, including the end points, the theorem gives the useful result that at points of discontinuity the value of the Fourier series for f takes the mean of the one sided limits of f itself at the discontinuous point. Given that the Fourier series is a continuous function (assuming the series to be uniformly convergent) representing f at this point of discontinuity this is the best that we can expect. Dirichlet's theorem is not therefore surprising. The formal proof of the theorem can be found in graduate texts such as Pinkus and Zafrany (1997) and depends on careful application of the Riemann–Lebesgue lemma and Bessel's inequality. Since f is periodic of period 2π, $f(\pi) = f(-\pi)$ and the last part of the theorem is seen to be nothing special, merely a re-statement that the Fourier series takes the mean of the one sided limits of f at discontinuous points.

We now state the basic theorem that enables piecewise continuous functions to be able to be expressed as Fourier series. The linear space notation is that used in Appendix C to which you are referred for more details.

Theorem 4.6 *The sequence of functions*

$$\left\{\frac{1}{\sqrt{2}}, \sin(x), \cos(x), \sin(2x), \cos(2x), \ldots\right\}$$

form an infinite orthonormal sequence in the space of all piecewise continuous functions on the interval $[-\pi, \pi]$ where the inner product $\langle f, g \rangle$ is defined by

$$\langle f, g \rangle = \frac{1}{\pi} \int_{-\pi}^{\pi} f \bar{g} \, dx$$

the overbar denoting complex conjugate.

Proof First we have to establish that $\langle f, g \rangle$ is indeed an inner product over the space of all piecewise continuous functions on the interval $[-\pi, \pi]$. The integral

$$\int_{-\pi}^{\pi} f \bar{g} \, dx$$

certainly exists. As f and $\bar{g}$ are piecewise continuous, so is the product $f\bar{g}$ and hence it is (Riemann) integrable. From elementary properties of integration it is easy to deduce that the space of all piecewise continuous functions is indeed an inner product space. There are no surprises. 0 and 1 are the additive and multiplicative identities, $-f$ is the additive inverse and the rules of algebra ensure associativity, distributivity and commutativity. We do, however spend some time establishing that the set

$$\left\{\frac{1}{\sqrt{2}}, \sin(x), \cos(x), \sin(2x), \cos(2x), \ldots\right\}$$

is orthonormal. To do this, it will be sufficient to show that

$$\left\langle \frac{1}{\sqrt{2}}, \frac{1}{\sqrt{2}} \right\rangle = 1, \quad \langle \sin(nx), \sin(nx) \rangle = 1,$$

$$\langle \cos(nx), \cos(nx) \rangle = 1, \quad \left\langle \frac{1}{\sqrt{2}}, \sin(nx) \right\rangle = 0,$$

$$\left\langle \frac{1}{\sqrt{2}}, \cos(nx) \right\rangle = 0, \quad \langle \cos(mx), \sin(nx) \rangle = 0,$$

$$\langle \cos(mx), \sin(nx) \rangle = 0, \quad \langle \sin(mx), \sin(nx) \rangle = 0,$$

with $m \neq n; m, n = 1, 2, \ldots$. Time spent on this is time well spent as orthonormality lies behind most of the important properties of Fourier series. For this, we do not use short cuts.

$$\left\langle \frac{1}{\sqrt{2}}, \frac{1}{\sqrt{2}} \right\rangle = \frac{1}{\pi} \int_{-\pi}^{\pi} \frac{1}{2} dx = 1 \text{ trivially}$$

$$\langle \sin(nx), \sin(nx) \rangle = \frac{1}{\pi} \int_{-\pi}^{\pi} \sin^2(nx) dx$$

$$= \frac{1}{2\pi} \int_{-\pi}^{\pi} (1 - \cos(2nx))dx = 1 \text{ for all } n$$

$$\langle \cos(nx), \cos(nx) \rangle = \frac{1}{\pi} \int_{-\pi}^{\pi} \cos^2(nx)dx$$

$$= \frac{1}{2\pi} \int_{-\pi}^{\pi} (1 + \cos(2nx))dx = 1 \text{ for all } n$$

$$\left\langle \frac{1}{\sqrt{2}}, \cos(nx) \right\rangle = \frac{1}{\pi} \int_{-\pi}^{\pi} \frac{1}{\sqrt{2}} \cos(nx)dx$$

$$= \frac{1}{\pi\sqrt{2}} \left[\frac{1}{n} \sin(nx) \right]_{-\pi}^{\pi} = 0 \text{ for all } n$$

$$\left\langle \frac{1}{\sqrt{2}}, \sin(nx) \right\rangle = \frac{1}{\pi} \int_{-\pi}^{\pi} \frac{1}{\sqrt{2}} \sin(nx)dx$$

$$= \frac{1}{\pi\sqrt{2}} \left[-\frac{1}{n} \cos(nx) \right]_{-\pi}^{\pi}$$

$$= \frac{1}{n\pi\sqrt{2}} ((-1)^n - (-1)^n) = 0 \text{ for all } n$$

$$\langle \cos(mx), \sin(nx) \rangle = \frac{1}{\pi} \int_{-\pi}^{\pi} \cos(mx) \sin(nx)dx$$

$$= \frac{1}{2\pi} \int_{-\pi}^{\pi} (\sin((m+n)x) + \sin((m-n)x)) \, dx$$

$$= \frac{1}{2\pi} \left[-\frac{\cos((m+n)x)}{m+n} - \frac{\cos((m-n)x)}{m-n} \right]_{-\pi}^{\pi} = 0, (m \neq n)$$

since the function in the square bracket is the same at both $-\pi$ and π. If $m = n$, $\sin((m - n)x) = 0$ but otherwise the arguments go through unchanged and $\langle \cos(mx), \sin(mx) \rangle = 0$, $m, n = 1, 2 \ldots$. Now

$$\langle \cos(mx), \cos(nx) \rangle = \frac{1}{\pi} \int_{-\pi}^{\pi} \cos(mx) \cos(nx)dx$$

$$= \frac{1}{2\pi} \int_{-\pi}^{\pi} (\cos((m+n)x) + \cos((m-n)x)) \, dx$$

$$= \frac{1}{2\pi} \left[\frac{\sin((m+n)x)}{m+n} + \frac{\sin((m-n)x)}{m-n} \right]_{-\pi}^{\pi} \quad \text{as } m \neq n$$

$$= 0 \text{ as all functions are zero at both limits.}$$

Finally,

$$\langle \sin(mx), \sin(nx) \rangle = \frac{1}{\pi} \int_{-\pi}^{\pi} \sin(mx) \sin(nx)dx$$

$$= \frac{1}{2\pi} \int_{-\pi}^{\pi} (\cos((m-n)x) - \cos((m+n)x)) \, dx$$

$= 0$ similarly to the previous result.

Hence the theorem is firmly established.

□

We have in the above theorem shown that the sequence

$$\left\{ \frac{1}{\sqrt{2}}, \sin(x), \cos(x), \sin(2x), \cos(2x), \ldots \right\}$$

is orthogonal. It is in fact also true that this sequence forms a basis (an *orthonormal* basis) for the space of piecewise continuous functions in the interval $[-\pi, \pi]$. This and other aspects of the theory of linear spaces, an outline of which is given in Appendix C thus ensures that an arbitrary element of the linear space of piecewise continuous functions can be expressed as a linear combination of the elements of this sequence, i.e.

$$f(x) \sim \frac{a_0}{\sqrt{2}} + a_1 \cos(x) + a_2 \cos(2x) + \cdots + a_n \cos(nx) + \cdots$$
$$+ b_1 \sin(x) + b_2 \sin(2x) + \cdots + b_n \sin(nx) + \cdots$$

so

$$f(x) \sim \frac{a_0}{\sqrt{2}} + \sum_{n=1}^{\infty} (a_n \cos(nx) + b_n \sin(nx)) \quad -\pi < x < \pi \qquad (4.1)$$

with the tilde being interpreted as follows. At points of discontinuity, the left hand side is the mean of the two one sided limits as dictated by Dirichlet's theorem. At points where the function is continuous, the right-hand side converges to $f(x)$ and the tilde means equals. This is the "standard" Fourier series expansion for $f(x)$ in the range $-\pi < x < \pi$. Since the right hand side is composed entirely of periodic functions, period 2π, it is necessary that

$$f(x) = f(x + 2N\pi) \quad N = 0, \pm 1, \pm 2, \ldots.$$

The authors of engineering texts are happy to start with Equation 4.1, then by multiplying through by $\sin(nx)$ (say) and integrating term by term between $-\pi$ and π, all but the b_n on the right disappears. This gives

$$b_n = \frac{1}{\pi} \int_{-\pi}^{\pi} f(x) \sin(nx) dx.$$

Similarly

$$a_n = \frac{1}{\pi} \int_{-\pi}^{\pi} f(x) \cos(nx) dx.$$

Of course, this gives the correct results, but questions about the legality or otherwise of dealing with and manipulating infinite series remain. In the context of linear spaces we can immediately write

$$f(x) = (a_0, a_1, b_1, \ldots, a_n, b_n, \ldots)$$

is a vector expressed in terms of the orthonormal basis

$$e = (e_0, e_{\alpha_1}, e_{\beta_1}, \ldots, e_{\alpha_n}, e_{\beta_n}, \ldots)$$

$$e_0 = \frac{1}{\sqrt{2}}, e_{\alpha_n} = \cos(nx), \quad e_{\beta_n} = \sin(nx),$$

so

$$\langle f, e_{\alpha_n} \rangle = \frac{1}{\pi} \int_{-\pi}^{\pi} f(x) \cos(nx) dx$$

and

$$\langle f, e_{\beta_n} \rangle = \frac{1}{\pi} \int_{-\pi}^{\pi} f(x) \sin(nx) dx.$$

The series

$$f = \sum_{k=0}^{\infty} \langle f, e_k \rangle e_k$$

where e_k, $k = 0, \ldots\ldots$ is a renumbering of the basis vectors. This is the standard expansion of f in terms of the orthonormal basis and is the Fourier series for f. Invoking the linear space theory therefore helps us understand how it is possible to express any function piecewise continuous in $[-\pi, \pi]$ as the series expansion (4.1),

$$f(x) \sim \frac{a_0}{\sqrt{2}} + \sum_{n=1}^{\infty} (a_n \cos(nx) + b_n \sin(nx)) \quad -\pi < x < \pi$$

where

$$a_n = \frac{1}{\pi} \int_{-\pi}^{\pi} f(x) \cos(nx) dx,$$

and

$$b_n = \frac{1}{\pi} \int_{-\pi}^{\pi} f(x) \sin(nx) dx, \; n = 0, 1, 2, \ldots.$$

and remembering to interpret correctly the left-hand side at discontinuities to be in line with Dirichlet's theorem. It is also now clear why a_0 is multiplied by $1/\sqrt{2}$ in all that has gone before, it is to ensure orthonormality. Unfortunately books differ as to where the factor goes. Some use $a_0 = 1$ with the factor $1/2$ in a separate integral for a_0. This should not done here as it contravenes the definition of orthonormality which is offensive to pure mathematicians everywhere. However it is standard practice to combine the factor $1/\sqrt{2}$ in the coefficient with the same factor that occurs when computing a_0 from $f(x)$ through the orthogonality relationships. The upshot of this combination is the "standard" Fourier series which is adopted from here on:

$$f(x) \sim \frac{1}{2} a_0 + \sum_{n=1}^{\infty} (a_n \cos(nx) + b_n \sin(nx)) \quad -\pi < x < \pi$$

where

$$a_n = \frac{1}{\pi} \int_{-\pi}^{\pi} f(x) \cos(nx)dx,$$

and

$$b_n = \frac{1}{\pi} \int_{-\pi}^{\pi} f(x) \sin(nx)dx, \ n = 0, 1, 2, \ldots.$$

We are now ready to do some practical examples. There is good news for those who perhaps are a little impatient with all this theory. It is not at all necessary to understand about linear space theory in order to calculate Fourier series. The earlier theory gives the framework in which Fourier series operate as well as enabling us to give decisive answers to key questions that can arise in awkward or controversial cases, for example if the existence or uniqueness of a particular Fourier series is in question. The first example is not controversial.

Example 4.7 *Determine the Fourier series for the function*

$$f(x) = 2x + 1 \quad -\pi < x < \pi$$
$$f(x) = f(x + 2\pi) \ x \in \mathbf{R}$$

where Theorem 4.5 applies at the end points.

Solution As $f(x)$ is obviously piecewise continuous in $[-\pi, \pi]$, in fact the only discontinuities occurring at the end points, we simply use the formulae

$$a_n = \frac{1}{\pi} \int_{-\pi}^{\pi} f(x) \cos(nx)dx,$$

and

$$b_n = \frac{1}{\pi} \int_{-\pi}^{\pi} f(x) \sin(nx)dx$$

to determine the Fourier coefficients. Now,

$$
\begin{aligned}
a_n &= \frac{1}{\pi} \int_{-\pi}^{\pi} (2x + 1) \cos(nx)dx \\
&= \frac{1}{\pi} \left[\frac{(2x + 1)}{n} \sin(nx) \right]_{-\pi}^{\pi} - \frac{1}{\pi} \int_{-\pi}^{\pi} \frac{2}{n} \sin(nx)dx \quad (n \neq 0) \\
&= 0 + \frac{1}{\pi} \frac{2}{n^2} [\cos(nx)]_{-\pi}^{\pi} = 0
\end{aligned}
$$

since $\cos(n\pi) = (-1)^n$. If $n = 0$ then

$$
\begin{aligned}
a_0 &= \frac{1}{\pi} \int_{-\pi}^{\pi} (2x + 1)dx \\
&= \frac{1}{\pi} [x^2 + x]_{-\pi}^{\pi} = 2
\end{aligned}
$$

$$b_n = \frac{1}{\pi} \int_{-\pi}^{\pi} (2x+1) \sin(nx) dx$$

$$= \frac{1}{\pi} \left[-\frac{(2x+1)}{n} \cos(nx) \right]_{-\pi}^{\pi} + \frac{1}{\pi} \int_{-\pi}^{\pi} \frac{2}{n} \cos(nx) dx$$

$$= \frac{1}{\pi} \left[-\frac{(2\pi+1)}{n} + \frac{(-2\pi+1)}{n} \right] (-1)^n$$

$$= -\frac{4}{n}(-1)^n.$$

Hence

$$f(x) \sim 1 - 4 \sum_{n=1}^{\infty} \frac{(-1)^n}{n} \sin(nx), \quad -\pi < x < \pi.$$

From this series, we can deduce the Fourier series for x as follows:-

we have $2x + 1 \sim 1 - 4 \sum_{n=1}^{\infty} \frac{(-1)^n}{n} \sin(nx), \quad x \in [-\pi, \pi]$

so

$$x \sim 2 \sum_{n=1}^{\infty} \frac{(-1)^{n+1}}{n} \sin(nx)$$

gives the Fourier series for x in $[-\pi, \pi]$.

Thus the Fourier series for the general straight line $y = mx + c$ in $[-\pi, \pi]$ must be

$$y \sim c - 2m \sum_{n=1}^{\infty} \frac{(-1)^n}{n} \sin(nx).$$

The explanation for the lack of cosine terms in these Fourier series follows later after the discussion of even and odd functions. Here is a slightly more involved example.

Example 4.8 *Find the Fourier series for the function*

$$f(x) = e^x, \quad -\pi < x < \pi$$

$$f(x + 2\pi) = f(x), \quad x \in \mathbf{R}$$

where Theorem 4.5 applies at the end points.

Solution This problem is best tackled by using the power of complex numbers. We start with the two standard formulae:

$$a_n = \frac{1}{\pi} \int_{-\pi}^{\pi} e^x \cos(nx) dx$$

and

$$b_n = \frac{1}{\pi} \int_{-\pi}^{\pi} e^x \sin(nx) dx$$

and form the sum $a_n + ib_n$ where $i = \sqrt{-1}$. The integration is then quite straightforward

$$
\begin{aligned}
a_n + ib_n &= \frac{1}{\pi} \int_{-\pi}^{\pi} e^{x+inx} dx \\
&= \frac{1}{\pi(1+in)} [e^{x+inx}]_{-\pi}^{\pi} \\
&= \frac{1}{\pi(1+in)} [e^{(1+in)\pi} - e^{-(1+in)\pi}] \\
&= \frac{(-1)^n}{\pi(1+in)} (e^{\pi} - e^{-\pi}) \text{ since } e^{in\pi} = (-1)^n \\
&= \frac{2(-1)^n \sinh \pi}{\pi(1+n^2)} (1 - in).
\end{aligned}
$$

Hence, taking real and imaginary parts we obtain

$$
a_n = \frac{2 \sinh(\pi)}{\pi(1+n^2)}, \quad b_n = -\frac{2n(-1)^n \sinh(\pi)}{\pi(1+n^2)}.
$$

In this example a_0 is given by

$$
a_0 = \frac{2}{\pi} \sinh(\pi),
$$

hence giving the Fourier series as

$$
f(x) = \frac{\sinh(\pi)}{\pi} + \frac{2}{\pi} \sinh(\pi) \sum_{n=1}^{\infty} \frac{(-1)^n}{1+n^2} (\cos(nx) - n \sin(nx)), \quad -\pi < x < \pi.
$$

Let us take this opportunity to make use of this series to find the values of some infinite series. The above series is certainly valid for $x = 0$ so inserting this value into both sides of the above equation and noting that $f(0)(= e^0) = 1$ gives

$$
1 = \frac{\sinh(\pi)}{\pi} + \frac{2}{\pi} \sum_{n=1}^{\infty} \frac{(-1)^n}{1+n^2}.
$$

Thus

$$
\sum_{n=1}^{\infty} \frac{(-1)^n}{1+n^2} = \frac{\pi}{2} \operatorname{cosech}(\pi) - \frac{1}{2}
$$

and

$$
\sum_{n=-\infty}^{\infty} \frac{(-1)^n}{1+n^2} = 2 \sum_{n=1}^{\infty} \frac{(-1)^n}{1+n^2} + 1 = \pi \operatorname{cosech}(\pi).
$$

Further, there is the opportunity here to use Dirichlet's theorem. Putting $x = \pi$ into the Fourier series for e^x is not strictly legal. However, Dirichlet's theorem states that the value of the series at this discontinuity is the mean of the one sided limits either side which is $\frac{1}{2}(e^{-\pi} + e^{\pi}) = \cosh \pi$. The Fourier series evaluated at $x = \pi$ is

$$\frac{\sinh(\pi)}{\pi} + \frac{2}{\pi} \sinh(\pi) \sum_{n=1}^{\infty} \frac{1}{1 + n^2}$$

as $\cos(n\pi) = (-1)^n$ and $\sin(n\pi) = 0$ for all integers n. The value of $f(x)$ at $x = \pi$ is taken to be that dictated by Dirichlet's theorem, viz. $\cosh(\pi)$. We therefore equate these expressions and deduce the series

$$\sum_{n=1}^{\infty} \frac{1}{1 + n^2} = \frac{\pi}{2} \coth(\pi) - \frac{1}{2}$$

and

$$\sum_{n=-\infty}^{\infty} \frac{1}{1 + n^2} = 2 \sum_{n=1}^{\infty} \frac{1}{1 + n^2} + 1 = \pi \coth(\pi).$$

Having seen the general method of finding Fourier series, we are ready to remove the restriction that all functions have to be of period 2π and defined in the range $[-\pi, \pi]$. The most straightforward way of generalising to Fourier series of any period is to effect the transformation $x \to \pi x/l$ where l is assigned by us. Thus if $x \in [-\pi, \pi]$, $\pi x/l \in [-l, l]$. Since $\cos(\pi x/l)$ and $\sin(\pi x/l)$ have period $2l$ the Fourier series valid in $[-l, l]$ takes the form

$$f(x) \sim \frac{1}{2}a_0 + \sum_{n=1}^{\infty} \left(a_n \cos\left(\frac{n\pi x}{l}\right) + b_n \sin\left(\frac{n\pi x}{l}\right) \right), \quad -l < x < l,$$

where

$$a_n = \frac{1}{l} \int_{-l}^{l} f(x) \cos\left(\frac{n\pi x}{l}\right) dx \qquad (4.2)$$

and

$$b_n = \frac{1}{l} \int_{-l}^{l} f(x) \sin\left(\frac{n\pi x}{l}\right) dx. \qquad (4.3)$$

The examples of finding this kind of Fourier series are not remarkable, they just contain (in general) messier algebra! Here is just one example.

Example 4.9 *Determine the Fourier Series of the function*

$$f(x) = |x|, \quad -3 \le x \le 3,$$
$$f(x) = f(x + 6).$$

Solution The function $f(x) = |x|$ is continuous, therefore we can use Equations 4.2 and 4.3 to generate the Fourier coefficients. First of all

$$a_0 = \frac{1}{3} \int_{-3}^{3} |x| dx = 3.$$

Secondly,

$$a_n = \frac{1}{3} \int_{-3}^{3} |x| \cos\left(\frac{n\pi x}{3}\right) dx$$

$$= \frac{1}{3} \int_{0}^{3} x \cos\left(\frac{n\pi x}{3}\right) dx$$

$$= \frac{2}{3} \left[\frac{3x}{n\pi} \sin\left(\frac{n\pi x}{3}\right)\right]_0^3 - \frac{2}{3} \int_0^3 \frac{3}{n\pi} \sin\left(\frac{n\pi x}{3}\right) dx$$

$$= 0 + \frac{2}{n\pi} \left[\frac{3}{n\pi} \cos\left(\frac{n\pi x}{3}\right)\right]_0^3$$

so $a_n = \frac{6}{(n\pi)^2}[-1 + (-1)^n]$.

This is zero if n is even and $-12/(n\pi)^2$ if n is odd. Hence

$$a_{2k+1} = -\frac{12}{(2k+1)^2 \pi^2}, \quad k = 0, 1, 2, \ldots$$

Similarly,

$$b_n = \frac{1}{3} \int_{-3}^{3} |x| \sin\left(\frac{n\pi x}{3}\right) dx$$

$$= -\frac{1}{3} \int_{-3}^{0} x \sin\left(\frac{n\pi x}{3}\right) dx + \frac{1}{3} \int_{0}^{3} x \sin\left(\frac{n\pi x}{3}\right) dx$$

$$= 0 \text{ for all } n.$$

Hence

$$f(x) = \frac{3}{2} - \frac{12}{\pi^2} \sum_{n=0}^{\infty} \frac{1}{(2n-1)^2} \cos\left[\frac{2n-1}{3}\pi x\right], \quad 0 \le x \le 3,$$

note the equality; $f(x)$ is continuous on $[0, 3]$ hence there is no discontinuity at the end points and no need to invoke Dirichlet's theorem.

4.3 Odd and Even Functions

This topic is often met at a very elementary level at schools in the context of how graphs of functions look on the page in terms of symmetries. However, here we give formal definitions and, more importantly, see how the identification of oddness or evenness in functions literally halves the amount of work required in finding the Fourier series.

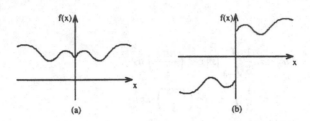

Figure 4.2: **a** An even function, **b** An odd function

Definition 4.10 *A function $f(x)$ is termed even with respect to the value a if*

$$f(a + x) = f(a - x)$$

for all values of x.

Definition 4.11 *A function $f(x)$ is termed odd with respect to the value a if*

$$f(a + x) = -f(a - x)$$

for all values of x.

The usual expressions "$f(x)$ is an even function" and "$f(x)$ is an odd function" means that $a = 0$ has been assumed. i.e. $f(x) = f(-x)$ means f is an even function and $f(x) = -f(-x)$ means f is an odd function. Well known even functions are :-

$$|x|, x^2, \cos(x).$$

Well known odd functions are

$$x, \sin(x), \tan(x).$$

An even function of x, plotted on the (x, y) plane, is symmetric about the y axis. An odd function of x drawn on the same axes is anti-symmetric (see Figure 4.2). The important consequence of the essential properties of these functions is that the Fourier series of an even function has to consist entirely of even functions and therefore has no sine terms. Similarly, the Fourier series of an odd function must consist entirely of odd functions, i.e. only sine terms.

Hence, given

$$f(x) = \frac{1}{2}a_0 + \sum_{n=1}^{\infty} (a_n \cos(nx) + b_n \sin(nx)) \quad -\pi < x < \pi$$

if $f(x)$ is even for all x then $b_n = 0$ for all n. If $f(x)$ is odd for all x then $a_n = 0$ for all n. We have already had one example of this. The function x is odd, and the Fourier series found after Example 4.7 is

$$x = 2 \sum_{n=1}^{\infty} \frac{(-1)^{n+1}}{n} \sin(nx), \quad -\pi < x < \pi$$

which has no even terms.

Example 4.12 *Determine the Fourier series for the function*

$$f(x) = x^2$$
$$f(x) = f(x + 2\pi), \quad -\pi \le x \le \pi.$$

Solution Since $x^2 = (-x)^2$, $f(x)$ is an even function. Thus the Fourier series consists solely of even functions which means $b_n = 0$ for all n. We therefore compute the a_n's as follows

$$a_0 = \frac{1}{\pi} \int_{-\pi}^{\pi} x^2 dx$$

$$= \frac{1}{3} \frac{1}{\pi} [x^3]_{-\pi}^{\pi} = \frac{2}{3} \pi^2$$

also $a_n = \dfrac{1}{\pi} \displaystyle\int_{-\pi}^{\pi} x^2 \cos(nx) dx \quad (n \ne 0)$

$$= \frac{1}{\pi} \left[\frac{x^2}{n} \sin(nx) \right]_{-\pi}^{\pi} - \frac{1}{\pi} \int_{-\pi}^{\pi} \frac{2x}{n} \sin(nx) dx$$

$$= 0 + \frac{1}{\pi} \left[\frac{2x}{n^2} \cos(nx) \right]_{-\pi}^{\pi} - \frac{2}{\pi n} \int_{-\pi}^{\pi} \frac{\cos(nx)}{n} dx$$

$$= \frac{4}{n^2} (-1)^n$$

so

$$f(x) = \frac{\pi^2}{3} + 4 \sum_{n=1}^{\infty} \frac{(-1)^n}{n^2} \cos nx \quad -\pi \le x < \pi.$$

This last example leads to some further insights.

If we let $x = 0$ in the Fourier series just obtained, the right-hand side is

$$\frac{\pi^2}{3} + 4 \sum_{n=1}^{\infty} \frac{(-1)^n}{n^2}$$

hence

$$0 = \frac{\pi^2}{3} + 4 \left(-1 + \frac{1}{2^2} - \frac{1}{3^2} + \cdots \right)$$

so

$$1 - \frac{1}{2^2} + \frac{1}{3^2} - \cdots = \frac{\pi^2}{12}.$$

This is interesting in itself as this series is not easy to sum. However it is also possible to put $x = \pi$ and obtain

$$\pi^2 = \frac{\pi^2}{3} + 4 \sum_{n=1}^{\infty} \frac{1}{n^2}$$

i.e.

$$\sum_{n=1}^{\infty} \frac{1}{n^2} = \frac{\pi^2}{6}.$$

Note that even periodic functions are continuous, whereas odd periodic functions are, using Dirichlet's theorem, zero at the end points. We shall utilise the properties of odd and even functions from time to time usually in order to simplify matters and reduce the algebra. Another tool that helps in this respect is the complex form of the Fourier series which is derived next.

4.4   Complex Fourier Series

Given that a Fourier series has the general form

$$f(x) \sim \frac{1}{2}a_0 + \sum_{n=1}^{\infty} (a_n \cos(nx) + b_n \sin(nx)), \quad -\pi < x < \pi,$$

we can write

$$\cos(nx) = \frac{1}{2}(e^{inx} + e^{-inx})$$

and

$$\sin(nx) = \frac{1}{2i}(e^{inx} - e^{-inx}).$$

If these equations are inserted into Equation 4.1 then we obtain

$$f(x) \sim \sum_{n=-\infty}^{\infty} c_n e^{inx}$$

where

$$c_n = \frac{1}{2}(a_n - ib_n), \quad c_{-n} = \frac{1}{2}(a_n + ib_n), \quad c_0 = \frac{1}{2}a_0, \quad n = 1, 2, \ldots.$$

Using the integrals for a_n and b_n we get

$$c_n = \frac{1}{2\pi} \int_{-\pi}^{\pi} f(x)e^{-inx}dx \text{ and } c_{-n} = \frac{1}{2\pi} \int_{-\pi}^{\pi} f(x)e^{inx}dx.$$

This is called the complex form of the Fourier series and can be useful for the computation of certain types of Fourier series. More importantly perhaps, it enables the step to Fourier Transforms to be made (Chapter 6) which not only unites this chapter and its subject, Fourier Series, to the earlier parts of the book, Laplace Transforms, but leads naturally to applications to the field of signal processing which is of great interest to many electrical engineers.

Example 4.13 *Find the Fourier series for the function*

$$f(t) = t^2 + t, \quad -\pi \leq t \leq \pi,$$
$$f(t) = f(t + 2\pi).$$

Solution We could go ahead and find the Fourier series in the usual way. However it is far easier to use the complex form but in a tailor-made way as follows. Given

$$a_n = \frac{1}{\pi} \int_{-\pi}^{\pi} f(t) \cos(nt) dt$$

and

$$b_n = \frac{1}{\pi} \int_{-\pi}^{\pi} f(t) \sin(nt) dt$$

we define

$$d_n = \frac{1}{\pi} \int_{-\pi}^{\pi} f(t) e^{int} dt = a_n + i b_n$$

so that

$$
\begin{aligned}
d_n &= \frac{1}{\pi} \int_{-\pi}^{\pi} (t^2 + t) e^{int} dt \\
&= \frac{1}{\pi} \left[\frac{t^2 + t}{in} e^{int} \right]_{-\pi}^{\pi} - \frac{1}{\pi} \int_{-\pi}^{\pi} \frac{2t + 1}{in} e^{int} dt \\
&= \frac{1}{\pi} \left[\frac{t^2 + t}{in} e^{int} - \frac{2t + 1}{(in)^2} e^{int} + \frac{2}{(in)^3} e^{int} \right]_{-\pi}^{\pi} \\
&= \frac{1}{\pi} \left[\frac{\pi^2 + \pi}{in} - \frac{\pi^2 - \pi}{in} + \frac{2\pi + 1}{n^2} - \frac{-2\pi + 1}{n^2} \right] (-1)^n \\
&= (-1)^n \left(\frac{4}{n^2} + \frac{2}{in} \right)
\end{aligned}
$$

so

$$a_n = \frac{4}{n^2} (-1)^n, \quad b_n = \frac{2}{n} (-1)^{n+1}$$

and

$$a_0 = \frac{1}{\pi} \int_{-\pi}^{\pi} (t^2 + t) dt = \frac{2}{3} \pi^2.$$

These can be checked by direct computation. The Fourier series is thus, assuming convergence of the right hand side,

$$f(t) \sim \frac{2}{3}\pi^2 + \sum_{n=1}^{\infty} (-1)^n \left(\frac{4\cos(nt)}{n^2} - \frac{2\sin(nt)}{n} \right) \qquad -\pi < t < \pi.$$

We shall use this last example to make a point that should be obvious at least with hindsight! In Example 4.7 we deduced the result

$$f(x) = 2x + 1 \sim 1 - 4 \sum_{n=1}^{\infty} \frac{(-1)^n}{n} \sin(nx), \quad -\pi \leq x \leq \pi.$$

The right hand side is pointwise convergent for all $x \in [-\pi, \pi]$. It is therefore legal (see Section 4.6) to integrate the above Fourier series term by term indefinitely. The left hand side becomes

$$\int_0^t (2x + 1)dx = t^2 + t.$$

The right hand side becomes

$$t - 4 \sum_{n=1}^{\infty} \frac{(-1)^n}{n^2} \cos(nt).$$

This result may seem to contradict the Fourier series just obtained, viz.

$$t^2 + t \sim \frac{2}{3}\pi^2 + \sum_{n=1}^{\infty} (-1)^n \left(\frac{4\cos(nt)}{n^2} - \frac{2\sin(nt)}{n} \right)$$

as the right hand sides are not the same. There is no contradiction however as

$$t - 4 \sum_{n=1}^{\infty} \frac{(-1)^n}{n^2} \cos(nt)$$

is not a Fourier series due to the presence of the isolated t at the beginning.

From a practical point of view, it is useful to know just how many terms of a Fourier series need to be calculated before a reasonable approximation to the periodic function is obtained. The answer of course depends on the specific function, but to get an idea of the approximation process, consider the function

$$f(t) = f(t + 2), \quad 0 \le t \le 2$$

which formally has the Fourier series

$$f(t) = 3 - \frac{2}{\pi} \sum_{n=1}^{\infty} \frac{\sin(\pi t)}{n}.$$

The sequence formed by the first seven partial sums of this Fourier series are shown superimposed in Figure 4.3. In this instance, it can be seen that there is quite a rapid convergence to the "saw-tooth" function $f(t) = t + 2$. Problems arise where there are rapid changes of gradient (at the corners) and in trying to approximate a vertical line via trigonometric series (which brings us back to Dirichlet's theorem). The overshoots at corners (Gibbs' phenomenon) and other problems (e.g. aliasing) are treated in depth in specialist texts. Here we concentrate on finding the series itself and now move on to some refinements.

4.5   Half Range Series

There is absolutely no problem explaining half range series in terms of the normed linear space theory of Section 4.1. However, we shall postpone this

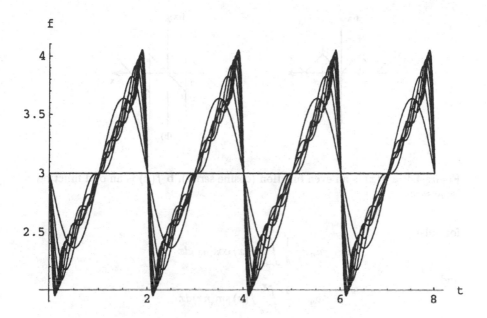

Figure 4.3: The first seven partial sums of the Fourier series for the function
$f(t) = t + 2$, $0 \leq t \leq 2$, $f(t) = f(t + 2)$ drawn in the range $0 \leq t \leq 8$.

until half range series have been explained and developed in terms of even and
odd functions. This is entirely natural, at least for the applied mathematician!

Half range series are, as the name implies, series defined over half of the
normal range. That is, for standard trigonometric Fourier series the function
$f(x)$ is defined only in $[0, \pi]$ instead of $[-\pi, \pi]$. The value that $f(x)$ takes in the
other half of the interval, $[-\pi, 0]$ is free to be defined. If we take

$$f(x) = f(-x)$$

that is $f(x)$ is even, then the Fourier series for $f(x)$ can be entirely expressed
in terms of even functions, i.e cosine terms. If on the other hand

$$f(x) = -f(-x)$$

then $f(x)$ is an odd function and the Fourier series is correspondingly odd and
consists only of sine terms. We are not defining the same function as two
different Fourier series, for $f(x)$ is different, at least over half the range (see
Figure 4.4). We are now ready to derive the half range series in detail. First
of all, let us determine the cosine series. Suppose

$$f(x) = f(x + 2\pi), \quad -\pi \leq x \leq \pi$$

and, additionally, $f(x) = f(-x)$ so that $f(x)$ is an even function. Since the

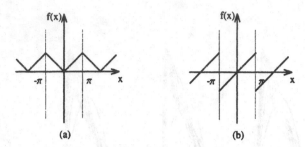

(a)   (b)

Figure 4.4: **a** $f(x)$ is an even function (cosine series), **b** $f(x)$ is an odd function (sine series)

formulae

$$a_n = \frac{1}{\pi} \int_{-\pi}^{\pi} f(x) \cos(nx) dx$$

and

$$b_n = \frac{1}{\pi} \int_{-\pi}^{\pi} f(x) \sin(nx) dx$$

have already been derived, we impose the condition that $f(x)$ is even. It is then easy to see that

$$a_n = \frac{2}{\pi} \int_0^{\pi} f(x) \cos nx dx$$

and $b_n = 0$. For odd functions, the formulae for a_n and b_n are $a_n = 0$ and

$$b_n = \frac{2}{\pi} \int_0^{\pi} f(x) \sin(nx) dx.$$

The following example brings these formulae to life.

Example 4.14 *Determine the half range sine and cosine series for the function*

$$f(t) = t^2 + t, \quad 0 \le t \le \pi.$$

Solution We have previously determined that, for $f(t) = t^2 + t$

$$a_n = \frac{4(-1)^n}{n^2}, \quad b_n = \frac{2}{n}(-1)^n.$$

For this function, the graph is displayed in Figure 4.5. If we wish $f(t)$ to be even, then $f(t) = f(-t)$ and so

$$a_n = \frac{2}{\pi} \int_0^{\pi} (t^2 + t) \cos(nt) dt, \quad b_n = 0.$$

On the other hand, if we wish $f(t)$ to be odd, we require $f(t) = -f(-t)$, where

$$b'_n = \frac{2}{\pi} \int_0^{\pi} (t^2 + t) \sin(nt) dt, \quad a'_n = 0.$$

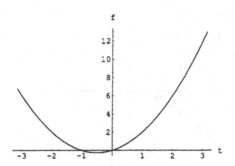

Figure 4.5: The graph $f(t) = t^2 + t$, shown for the entire range $-\pi \leq t \leq \pi$ to emphasise that it might be periodic but it is neither odd nor even

Both of these series can be calculated quickly using the same trick as used to determine the whole Fourier series, namely, let

$$k = a_n + ib'_n = \frac{2}{\pi} \int_0^\pi (t^2 + t)e^{int}dt.$$

We evaluate this carefully using integration by parts and show the details.

$$
\begin{aligned}
k &= \frac{2}{\pi} \int_0^\pi (t^2 + t)e^{int}dt \\
&= \frac{2}{\pi} \left[\frac{t^2 + t}{in}e^{int} \right]_0^\pi - \frac{2}{\pi} \int_0^\pi \frac{2t + 1}{in}e^{int}dt \\
&= \frac{2}{\pi} \left[\frac{t^2 + t}{in}e^{int} - \frac{2t + 1}{(in)^2}e^{int} + \frac{2}{(in)^3}e^{int} \right]_0^\pi \\
&= \frac{2}{\pi} \left[-\frac{\pi^2 + \pi}{in} + \frac{2\pi + 1}{n^2} + \frac{2i}{n^3} \right](-1)^n \\
&\quad - \frac{2}{\pi} \left[-\frac{1}{n^2} + \frac{2i}{n^3} \right]
\end{aligned}
$$

from which

$$a_n = \frac{2}{\pi} \left[\frac{2\pi + 1}{n^2}(-1)^n + \frac{1}{n^2} \right]$$

and

$$b'_n = \frac{2}{\pi} \left[-\frac{(\pi^2 + \pi)}{n}(-1)^n + \frac{2}{n^3}((-1)^n - 1) \right].$$

The constant term a_0 is given by

$$a_0 = \frac{2}{\pi} \int_0^\pi (t^2 + t)dt$$

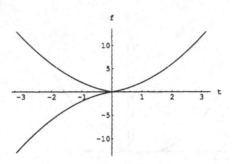

Figure 4.6: The function $f(t)$ displayed both as an even function and as an odd function

$$= \frac{1}{\pi} \left[\frac{1}{3}t^3 + \frac{1}{2}t^2 \right]_0^\pi$$

$$= \left[\frac{\pi^2}{3} + \frac{\pi}{2} \right]$$

so the constant term in the Fourier series $(a_0/2)$ is

$$\frac{1}{2} \left[\frac{\pi^2}{3} + \frac{\pi}{2} \right].$$

Therefore we have deduced the following two series: the *even* series for $t^2 + t$:-

$$\frac{1}{2} \left[\frac{\pi^2}{3} + \frac{\pi}{2} \right] + \frac{2}{\pi} \sum_{n=1}^{\infty} \left[\frac{2\pi + 1}{n^2}(-1)^n + \frac{1}{n^2} \right] \cos(nt), \quad x \in (0, \pi)$$

and the *odd* series for $t^2 + t$

$$\frac{2}{\pi} \sum_{n=1}^{\infty} \left[\frac{(\pi^2 + \pi)}{n}(-1)^{n+1} + \frac{2}{n^3}((-1)^n - 1) \right] \sin(nt), \quad x \in (0, \pi).$$

They are pictured in Figure 4.6. To frame this in terms of the theory of linear spaces, the originally chosen set of basis functions

$$\frac{1}{\sqrt{2}}, \cos(x), \sin(x), \cos(2x), \sin(2x), \ldots$$

is no longer a basis in the halved interval $[0, \pi]$. However the sequences

$$\frac{1}{\sqrt{2}}, \cos(x), \cos(2x), \ldots$$

and

$$\sin(x), \sin(2x), \sin(3x) \ldots$$

are, separately, both bases. Half range series are thus legitimate.

4.6   Properties of Fourier Series

In this section we shall be concerned with the integration and differentiation of Fourier series. Intuitively, it is the differentiation of Fourier series that poses more problems than integration. This is because differentiating $\cos(nx)$ or $\sin(nx)$ with respect to x gives $-n\sin(nx)$ or $n\cos(nx)$ which for large n are both larger in magnitude than the original terms. Integration on the other hand gives $\sin(nx)/n$ or $-\cos(nx)/n$, both smaller in magnitude. For those familiar with numerical analysis this comes as no surprise as numerical differentiation always needs more care than numerical integration which by comparison is safe. The following theorem covers the differentiation of Fourier series.

Theorem 4.15 *If f is continuous on $[-\pi, \pi]$ and piecewise differentiable in $(-\pi, \pi)$ which means that the derivative f' is piecewise continuous on $[-\pi, \pi]$, and if $f(x)$ has the Fourier series*

$$f(x) \sim \frac{1}{2}a_0 + \sum_{n=1}^{\infty} \{a_n \cos(nx) + b_n \sin(nx)\}$$

then the Fourier series of the derivative of $f(x)$ is given by

$$f'(x) \sim \sum_{n=1}^{\infty} \{-na_n \sin(nx) + nb_n \cos(nx)\}.$$

The proof of this theorem follows standard analysis and is not given here. The integration of a Fourier series poses less of a problem and can virtually always take place. The conditions for a function $f(x)$ to possess a Fourier series are similar to those required for integrability (piecewise continuity is sufficient), therefore integrating term by term can occur. A minor problem arises because the result is not necessarily another Fourier series. A term linear in x is produced by integrating the constant term whenever this is not zero. Formally, the following theorem covers the integration of Fourier series. It is not proved either, although a related more general result is derived a little later as a precursor to Parseval's theorem.

Theorem 4.16 *If f is piecewise continuous on the interval $[-\pi, \pi]$ and has the Fourier series*

$$f(x) \sim \frac{1}{2}a_0 + \sum_{n=1}^{\infty} \{a_n \cos(nx) + b_n \sin(nx)\}$$

then for each $x \in [-\pi, \pi]$,

$$\int_{-\pi}^{x} f(t)dt = \frac{1}{2}a_0(x + \pi) + \sum_{n=1}^{\infty} \left[\frac{a_n}{n} \sin(nx) - \frac{b_n}{n} (\cos(nx) - \cos(n\pi)) \right]$$

and the function on the right converges uniformly to the function on the left.

Let us discuss the details of differentiating and integrating Fourier series via the three series for the functions x^3, x^2 and x in the range $[-\pi, \pi]$. Formally, these three functions can be rendered periodic of period 2π by demanding that for each, $f(x) = f(x + 2\pi)$ and using Theorem 4.5 at the end points. The three Fourier series themselves can be derived using Equation 4.1 and are

$$x^3 \sim \sum_{n=1}^{\infty} (-1)^n \frac{2}{n^3} (6 - \pi^2 n^2) \sin(nx)$$

$$x^2 \sim \frac{\pi^2}{3} + \sum_{n=1}^{\infty} \frac{4(-1)^n}{n^2} \cos(nx)$$

$$x \sim \sum_{n=1}^{\infty} \frac{2}{n} (-1)^{n+1} \sin(nx)$$

all valid for $-\pi < x < \pi$. We state without proof the following facts about these three series. The series for x^2 is uniformly convergent. Neither the series for x nor that for x^3 are uniformly convergent. All the series are pointwise convergent. It is therefore legal to differentiate the series for x^2 but not either of the other two. All the series can be integrated. Let us perform the operations and verify these claims. It is certainly true that the term by term differentiation of the series

$$x^2 \sim \frac{\pi^2}{3} + \sum_{n=1}^{\infty} \frac{4(-1)^n}{n^2} \cos(nx)$$

gives

$$2x \sim \sum_{n=1}^{\infty} \frac{4}{n} (-1)^{n+1} \sin(nx)$$

which is the same as the Fourier series for x apart from the trivial factor of 2. Integrating a Fourier series term by term leads to the generation of an arbitrary constant. This can only be evaluated by the insertion of a particular value of x. To see how this works, let us integrate the series for x^2 term by term. The result is

$$\frac{x^3}{3} \sim \frac{\pi^2}{3} x + A + \sum_{n=1}^{\infty} \frac{4}{n^3} (-1)^n \sin(nx)$$

where A is an arbitrary constant. Putting $x = 0$ gives $A = 0$ and inserting the Fourier series for x in place of the x on the right hand side regains the Fourier series for x^3 already given. This integration of Fourier series is not always productive. Integrating the series for x term by term is not useful as there is no easy way of evaluating the arbitrary constant that is generated (unless one happens to know the value of some obscure series). Note also that blindly (and *illegally*) differentiating the series for x^3 or x term by term give nonsense in both cases. Engineers need to take note of this!

Let us now derive a more general result involving the integration of Fourier series. Suppose $F(t)$ is piecewise differentiable in the interval $(-\pi, \pi)$ and therefore continuous on the interval $[-\pi, \pi]$. Let $F(t)$ be represented by the Fourier series

$$F(t) \sim \frac{1}{2}A_0 + \sum_{n=1}^{\infty} (A_n \cos(nt) + B_n \sin(nt)), \quad -\pi < t < \pi,$$

and the usual periodicity

$$F(t) = F(t + 2\pi).$$

Now suppose that we can define another function $G(x)$ through the relationship

$$\int_{-\pi}^{t} G(x)dx = \frac{1}{2}a_0 t + F(t).$$

We then set ourselves the task of determining the Fourier series for $G(x)$. Using that $F(t)$ has a full range Fourier series we have

$$A_n = \frac{1}{\pi}\int_{-\pi}^{\pi} F(t)\cos(nt)dt, \text{ and } B_n = \frac{1}{\pi}\int_{-\pi}^{\pi} F(t)\sin(nt)dt.$$

By the fundamental theorem of the calculus we have that

$$F'(t) = G(t) - \frac{1}{2}a_0.$$

Now, evaluating the integral for A_n by parts once gives

$$A_n = -\frac{1}{n\pi}\int_{-\pi}^{\pi} F'(t)\sin(nt)dt,$$

and similarly,

$$B_n = \frac{1}{n\pi}\int_{-\pi}^{\pi} F'(t)\cos(nt)dt.$$

Hence, writing the Fourier series for $G(x)$ as

$$G(x) = \frac{1}{2}a_0 + \sum_{n=1}^{\infty} (a_n \cos(nx) + b_n \sin(nx)), \quad -\pi < x < \pi,$$

and comparing with

$$G(t) = \frac{1}{2}a_0 + F'(t)$$

gives

$$A_n = -\frac{b_n}{n} \text{ and } B_n = \frac{a_n}{n}.$$

We now need to consider the values of the Fourier series for $F(t)$ at the end points $t = -\pi$ and $t = \pi$. With

$$t = \pi, \quad F(\pi) = -\frac{1}{2}a_0\pi + \int_{-\pi}^{\pi} G(x)dx$$

and with

$$t = -\pi, \ F(-\pi) = \frac{1}{2}a_0\pi.$$

Since $F(t)$ is periodic of period 2π we have

$$F(-\pi) = F(\pi) = \frac{1}{2}a_0\pi.$$

Also

$$F(t) \sim \frac{1}{2}A_0 + \sum_{n=1}^{\infty} \left(-\frac{b_n}{n}\cos(nt) + \frac{a_n}{n}\sin(nt) \right).$$

Putting $t = \pi$ is legitimate in the limit as there is no jump discontinuity at this point (piecewise differentiability), and gives

$$\frac{1}{2}a_0\pi = \frac{1}{2}A_0 - \sum_{n=1}^{\infty} \frac{b_n}{n}\cos(n\pi) = \frac{1}{2}A_0 - \sum_{n=1}^{\infty}(-1)^n\frac{b_n}{n}.$$

We need to relate A_0 to a_0 and the b_n terms. To do this, we use the new form of $F(t)$ namely

$$F(t) \sim \frac{1}{2}a_0\pi + \sum_{n=1}^{\infty} \frac{1}{n}[b_n((-1)^n - \cos(nt)) + a_n\sin(nt)]$$

which is written in terms of the Fourier coefficients of $G(t)$, the integral of $F(t)$. To determine the form of this integral we note that

$$\int_{-\pi}^{t} G(x)dx = \frac{1}{2}a_0t + F(t)$$

$$= \frac{1}{2}a_0(\pi + t) + \sum_{n=1}^{\infty} \frac{1}{n}[b_n(-1)^n - b_n\cos(nt) + a_n\sin(nt)].$$

Also

$$\int_{-\pi}^{\xi} G(x)dx = \frac{1}{2}a_0(\pi + \xi) + \sum_{n=1}^{\infty} \frac{1}{n}[b_n(-1)^n - b_n\cos(n\xi) + a_n\sin(n\xi)]$$

so subtracting gives

$$\int_{\xi}^{t} G(x)dx = \frac{1}{2}a_0(t-\xi) + \sum_{n=1}^{\infty} \frac{1}{n}[b_n(\cos(n\xi) - \cos(nt)) + a_n(\sin(nt) - \sin(n\xi))]$$

which tells us the form of the integral of a Fourier series. In fact we alluded to this in Example 4.13. Here is an example where the ability to integrate a Fourier series term by term proves particularly useful.

Example 4.17 *Use the Fourier series*

$$x^2 = \frac{\pi^2}{3} + 4 \sum_{n=1}^{\infty} \frac{(-1)^n}{n^2} \cos(nx)$$

to deduce the value of the series

$$\sum_{k=1}^{\infty} \frac{(-1)^{k-1}}{(2k-1)^3}.$$

Solution Utilising the result just derived on the integration of Fourier series, we put $\xi = 0$ and $t = \pi/2$ so we can write

$$\int_0^{\frac{\pi}{2}} x^2 dx = \frac{\pi^2}{3} \left(\frac{\pi}{2} - 0 \right) + 4 \sum_{n=1}^{\infty} \frac{1}{n} \left[\frac{(-1)^n}{n^2} \sin \left(\frac{n\pi}{2} \right) \right]$$

so

$$\frac{\pi^3}{24} = \frac{\pi^3}{6} + 4 \sum_{n=1}^{\infty} \frac{(-1)^n}{n^3} \sin \left(\frac{n\pi}{2} \right).$$

Now, since $\sin(n\pi/2)$ takes the values $0, 1, 0, \ 1, \ldots$ for $n = 0, 1, 2, \ldots$ we immediately deduce that

$$\sum_{n=1}^{\infty} \frac{(-1)^n}{n^3} \sin \left(\frac{n\pi}{2} \right) = -\sum_{k=1}^{\infty} \frac{(-1)^{k-1}}{(2k-1)^3}$$

which gives

$$\sum_{n=1}^{\infty} \frac{(-1)^n}{n^3} \sin \left(\frac{n\pi}{2} \right) = \frac{1}{4} \left(\frac{\pi^3}{6} - \frac{\pi^3}{24} \right) = \frac{\pi^3}{32}.$$

Whence, putting $n = 2k - 1$ gives the result

$$\sum_{k=1}^{\infty} \frac{(-1)^{k-1}}{(2k-1)^3} = \frac{\pi^3}{32}.$$

The following theorem can now be deduced.

Theorem 4.18 *If $f(t)$ and $g(t)$ are continuous in $(-\pi, \pi)$ and provided*

$$\int_{-\pi}^{\pi} |f(t)|^2 dt < \infty \ \text{ and } \ \int_{-\pi}^{\pi} |g(t)|^2 dt < \infty,$$

if a_n, b_n are the Fourier coefficients of $f(t)$ and α_n, β_n those of $g(t)$, then

$$\int_{-\pi}^{\pi} f(t)g(t)dt = \frac{1}{2}\pi a_0 b_0 + \pi \sum_{n=1}^{\infty} (\alpha_n a_n + \beta_n b_n).$$

Proof Since

$$f(t) \sim \frac{a_0}{2} + \sum_{n=1}^{\infty} (a_n \cos(nt) + b_n \sin(nt))$$

and

$$g(t) \sim \frac{b_0}{2} + \sum_{n=1}^{\infty} (\alpha_n \cos(nt) + \beta_n \sin(nt))$$

we can write

$$f(t)g(t) \sim \frac{1}{2}a_0 g(t) + \sum_{n=1}^{\infty} (a_n g(t) \cos(nt) + b_n g(t) \sin(nt)).$$

Integrating this series from $-\pi$ to π gives

$$\int_{-\pi}^{\pi} f(t)g(t)dt = \frac{1}{2}a_0 \int_{-\pi}^{\pi} g(t)dt$$
$$+ \sum_{n=1}^{\infty} \left\{ a_n \int_{-\pi}^{\pi} g(t) \cos(nt)dt + b_n \int_{-\pi}^{\pi} g(t) \sin(nt)dt \right\}$$

provided the Fourier series for $f(t)$ is uniformly convergent, enabling the summation and integration operations to be interchanged. This follows from the Cauchy–Schwarz inequality (see Appendix C, Theorem C1) since

$$\int_{-\pi}^{\pi} |f(t)g(t)|dt \leq \left(\int_{-\pi}^{\pi} |f(t)|^2 dt \right)^{1/2} \left(\int_{-\pi}^{\pi} |g(t)|^2 dt \right)^{1/2} < \infty.$$

However, we know that

$$\frac{1}{\pi} \int_{-\pi}^{\pi} g(t) \cos(nt)dt = \alpha_n$$
$$\frac{1}{\pi} \int_{-\pi}^{\pi} g(t) \sin(nt)dt = \beta_n$$

and that

$$\alpha_0 = \frac{2}{\pi} \int_{-\pi}^{\pi} g(t)dt$$

so this implies

$$\int_{-\pi}^{\pi} f(t)g(t)dt = \frac{1}{2}\pi a_0 b_0 + \pi \sum_{n=1}^{\infty} (\alpha_n a_n + \beta_n b_n)$$

as required.

□

If we put $f(t) = g(t)$ in the above result, the following important theorem immediately follows.

Theorem 4.19 (Parseval) *If $f(t)$ is continuous in the range $(-\pi, \pi)$, is square integrable (i.e. $\int_{-\pi}^{\pi} [f(t)]^2 dt < \infty$) and has Fourier coefficients a_n, b_n then*

$$\int_{-\pi}^{\pi} [f(t)]^2 dt = 2\pi a_0^2 + \pi \sum_{n=1}^{\infty} (a_n^2 + b_n^2).$$

This is a useful result for mathematicians, but perhaps its most helpful attribute lies in its interpretation. The left hand side represents the mean square value of $f(t)$ (once it is divided by 2π). It can therefore be thought of in terms of energy if $f(t)$ represents a signal. What Parseval's theorem states therefore is that the energy of a signal expressed as a waveform is proportional to the sum of the squares of its Fourier coefficients. In Chapter 6 when Fourier Transforms are discussed, Parseval's theorem re-emerges in this practical context, perhaps in a more recognisable form. For now, let us content ourselves with a mathematical consequence of the theorem.

Example 4.20 *Given the Fourier series*

$$t^2 - \frac{\pi^2}{3} + 4 \sum_{n=1}^{\infty} \frac{(-1)^n}{n^2} \cos(nt)$$

deduce the value of

$$\sum_{n=1}^{\infty} \frac{1}{n^4}.$$

Solution Applying Parseval's theorem to this series, the left hand side becomes

$$\int_{-\pi}^{\pi} (t^2)^2 dt = \frac{2}{5}\pi^5.$$

The right hand side becomes

$$2\pi \left(\frac{\pi^2}{3}\right)^2 + \pi \sum_{n=1}^{\infty} \frac{16}{n^4}.$$

Equating these leads to

$$\frac{2}{5}\pi^5 = \frac{2}{9}\pi^5 + \pi \sum_{n=1}^{\infty} \frac{16}{n^4}$$

or

$$16 \sum_{n=1}^{\infty} \frac{1}{n^4} = \pi^4 \left(\frac{2}{5} - \frac{2}{9}\right).$$

Hence

$$\sum_{n=1}^{\infty} \frac{1}{n^4} = \frac{\pi^4}{90}.$$

4.7 Exercises

(Note: The first two exercises depend more on knowledge of Appendix C and may be left if desired.)

1. Use the Riemann–Lebesgue lemma to show that

$$\lim_{m \to \infty} \int_0^\pi g(t) \sin\left(m + \frac{1}{2}\right) t\, dt = 0,$$

   where $g(t)$ is piecewise continuous on the interval $[0, \pi]$.

2. Show that the functions $P_n(t)$ that satisfy the ordinary differential equation

$$(1 - t^2)\frac{d^2 P_n}{dt^2} - 2t\frac{dP_n}{dt} + n(n+1)P_n = 0, \quad n = 0, 1, \ldots$$

   are orthogonal in the interval $[-1, 1]$, where the inner product is defined by

$$\langle p, q \rangle = \int_{-1}^1 pq\, dt.$$

3. $f(t)$ is defined by

$$f(t) = \begin{cases} t & 0 \le t \le \frac{1}{2}\pi \\ \frac{1}{2}\pi & \frac{1}{2}\pi \le t \le \pi \\ \pi - \frac{1}{2}t & \pi \le t \le 2\pi. \end{cases}$$

   Sketch the graph of $f(t)$ and determine a Fourier series assuming $f(t) = f(t + 2\pi)$.

4. Determine the Fourier series for $f(x) = H(x)$, the Heaviside Unit Step Function, in the range $[-\pi, \pi]$, $f(x) = f(x + 2\pi)$. Hence find the value of the series

$$1 - \frac{1}{3} + \frac{1}{5} - \frac{1}{7} + \frac{1}{9} - \cdots.$$

5. Find the Fourier series of the function

$$f(x) = \begin{cases} \sin(\frac{1}{2}x) & 0 \le x \le \pi \\ -\sin(\frac{1}{2}x) & \pi < x \le 2\pi \end{cases}$$

   with $f(x) = f(x + 2\pi)$.

6. Determine the Fourier series for the function $f(x) = 1 - x^2$, $f(x) = f(x + 2\pi)$. Suggest possible values of $f(x)$ at $x = \pi$.

7. Deduce that the Fourier series for the function $f(x) = e^{ax}$, $-\pi < x < \pi$, a a real number is

$$\frac{\sinh(\pi a)}{\pi}\left\{\frac{1}{a} + 2\sum_{n=1}^\infty \frac{(-1)^n}{a^2 + n^2}\left(a\cos(nx) - n\sin(nx)\right)\right\}.$$

Hence find the values of the four series:

$$\sum_{n=1}^{\infty} \frac{(-1)^n}{a^2 + n^2}, \quad \sum_{n=-\infty}^{\infty} \frac{(-1)^n}{a^2 + n^2}, \quad \sum_{n=1}^{\infty} \frac{1}{a^2 + n^2}, \quad \sum_{n=-\infty}^{\infty} \frac{1}{a^2 + n^2}.$$

8. If

$$f(t) = \begin{cases} -t + e^t & -\pi \le t < 0 \\ t + e^t & 0 \le t < \pi \end{cases}$$

where $f(t) = f(t + 2\pi)$, sketch the graph of $f(t)$ for $-4\pi \le t \le 4\pi$ and obtain a Fourier series expansion for $f(t)$.

9. Find the Fourier series expansion of the function $f(t)$ where

$$f(t) = \begin{cases} \pi^2 & -\pi < t < 0 \\ (t - \pi)^2 & 0 \le t < \pi. \end{cases}$$

and $f(t) = f(t + 2\pi)$. Hence determine the values of the series

$$\sum_{n=1}^{\infty} \frac{1}{n^2} \quad \text{and} \quad \sum_{n=1}^{\infty} \frac{(-1)^{n+1}}{n^2}.$$

10. Determine the two Fourier half-range series for the function $f(t)$ defined in Exercise 9, and sketch the graphs of the function in both cases over the range $[-2\pi \le t \le 2\pi]$.

11. Given the half range sine series

$$t(\pi - t) = \frac{8}{\pi} \sum_{n=1}^{\infty} \frac{\sin(2n-1)t}{(2n-1)^3}, \quad 0 \le t \le \pi$$

use Parseval's Theorem to deduce the value of the series $\displaystyle\sum_{n=1}^{\infty} \frac{1}{(2n-1)^6}$.

Hence deduce the value of the series $\displaystyle\sum_{n=1}^{\infty} \frac{1}{n^6}$.

12. Deduce that the Fourier series for the function $f(x) = x^4$, $-\pi < x < \pi$, is

$$x^4 \sim \frac{\pi^4}{5} + \sum_{n=1}^{\infty} \frac{8(-1)^n}{n^4} (\pi^2 n^2 - 6) \cos(nx).$$

Explain why this series contains no sine terms. Use this series to find the value of the series

$$\sum_{n=1}^{\infty} \frac{(-1)^{n+1}}{n^4}$$

given that

$$\sum_{n=1}^{\infty} \frac{(-1)^{n+1}}{n^2} = \frac{\pi^2}{12}.$$

Assuming (correctly) that this Fourier series is uniformly convergent, use it to derive the Fourier series for x^3 over the range $-\pi < x < \pi$.

13. Given the Fourier series

$$x \sim \sum_{n=1}^{\infty} \frac{2}{n}(-1)^{n+1}\sin(nx), \quad -\pi < x < \pi,$$

integrate term by term to obtain the Fourier series for x^2, evaluating the constant of integration by integrating both sides over the range $[-\pi, \pi]$. Use the same integration technique on the Fourier series for x^4 given in the last exercise to deduce the Fourier series for x^5 over the range $-\pi < x < \pi$.

14. In an electrical circuit, the voltage is given by the "top-hat" function

$$V(t) = \begin{cases} 40 & 0 < t < 2 \\ 0 & 2 < t < 5. \end{cases}$$

Obtain the first five terms of the complex Fourier series for $V(t)$.

5

Partial Differential Equations

5.1 Introduction

In previous chapters, we have explained how ordinary differential equations can be solved using Laplace Transforms. In Chapter 4, Fourier series were introduced, and the important property that any reasonable function can be expressed as a Fourier series derived. In this chapter, these ideas are brought together, and the solution of certain types of partial differential equation using both Laplace Transforms and Fourier Series are explored. The study of the solution of partial differential equations (abbreviated PDEs) is a vast topic that it is neither possible nor appropriate to cover in a single chapter. There are many excellent texts (Sneddon (1957) and Williams (1980) to name but two) that have become standard. Here we shall only be interested in certain types of PDE that are amenable to solution by Laplace Transform.

Of course, to start with we will have to assume you know something about partial derivatives! If a function depends on more than one variable, then it is in general possible to differentiate it with respect to one of them provided all the others are held constant while doing so. Thus, for example, a function of three variables $f(x, y, z)$ (if differentiable in all three) will have three derivatives written

$$\frac{\partial f}{\partial x}, \frac{\partial f}{\partial y}, \text{ and } \frac{\partial f}{\partial z}.$$

The three definitions are straightforward and, hopefully, familiar.

$$\frac{\partial f}{\partial x} = \lim_{\Delta x \to 0} \left\{ \frac{f(x + \Delta x, y, z) - f(x, y, z)}{\Delta x} \right\}$$

y and z are held constant,

$$\frac{\partial f}{\partial y} = \lim_{\Delta y \to 0} \left\{ \frac{f(x, y + \Delta y, z) - f(x, y, z)}{\Delta y} \right\}$$

x and z are held constant, and

$$\frac{\partial f}{\partial z} = \lim_{\Delta z \to 0} \left\{ \frac{f(x, y, z + \Delta z) - f(x, y, z)}{\Delta z} \right\}$$

x and y are held constant. If all this is deeply unfamiliar, mysterious and a little terrifying, then a week or two with an elementary text on partial differentiation is recommended. It is an easy task to perform: simply differentiate with respect to one of the variables whilst holding the others constant. Also, it is easy to deduce that all the normal rules of differentiation apply as long it is remembered which variables are constant and which is the one to which the function is being differentiated. One example makes all this clear.

Example 5.1 *Find all first order partial derivatives of the functions (a) x^2yz, and (b) $x\sin(x + yz)$.*

Solution (a) The partial derivatives are as follows:-

$$\frac{\partial}{\partial x}(x^2yz) = 2xyz$$

$$\frac{\partial}{\partial y}(x^2yz) = x^2z$$

and

$$\frac{\partial}{\partial z}(x^2yz) = x^2y.$$

(b) The partial derivatives are as follows:-

$$\frac{\partial}{\partial x}(x\sin(x + yz)) = \sin(x + yz) + x\cos(x + yz)$$

which needs the product rule,

$$\frac{\partial}{\partial y}(x\sin(x + yz)) = xz\cos(x + yz)$$

and

$$\frac{\partial}{\partial z}(x\sin(x + yz)) = xy\cos(x + yz)$$

which do not.

There are chain rules for determining partial derivatives when f is a function of u, v and w which in turn are functions of x, y and z

$$u = u(x, y, z), \; v = v(x, y, z) \text{ and } w = w(x, y, z).$$

This is direct extension of the "function of a function" rule for single variable differentiation. There are other new features such as the Jacobian. We shall not pursue these here; instead the interested reader is referred to specialist texts such as Weinberger (1965) or Zauderer (1989).

5.2 Classification of Partial Differential Equations

In this book, we will principally be concerned with those partial differential equations that can be solved using Laplace Transforms, perhaps with the aid of Fourier Series. Thus we will eventually concentrate on second order PDEs of a particular type. However, in order to place these in context, we need to quickly review (or introduce for those readers new to this subject) the three different generic types of second order PDE.

The general second order PDE can be written

$$a_1 \frac{\partial^2 \phi}{\partial x^2} + b_1 \frac{\partial^2 \phi}{\partial x \partial y} + c_1 \frac{\partial^2 \phi}{\partial y^2} + d_1 \frac{\partial \phi}{\partial x} + e_1 \frac{\partial \phi}{\partial y} + f_1 \phi = g_1 \qquad (5.1)$$

where $a_1, b_1, c_1, d_1, e_1, f_1$ and g_1 are suitably well behaved functions of x and y. However, this is not a convenient form of the PDE for ϕ. By judicious use of Taylor's theorem and simple co-ordinate transformations it can be shown (e.g. Williams (1980), Chapter 3) that there are *three* basic types of linear second order partial differential equation. These standard types of PDE are termed hyperbolic, parabolic and elliptic following geometric analogies and are referred to as *canonical forms*. A crisp notation we introduce at this point is the *suffix derivative notation* whereby

$$\phi_x = \frac{\partial \phi}{\partial x}$$

$$\phi_y = \frac{\partial \phi}{\partial y}, \text{ etc.}$$

$$\phi_{xx} = \frac{\partial^2 \phi}{\partial x^2}$$

$$\phi_{yy} = \frac{\partial^2 \phi}{\partial y^2}$$

$$\phi_{xy} = \frac{\partial^2 \phi}{\partial x \partial y}, \text{ etc.}$$

This notation is very useful when writing large complicated expressions that involve partial derivatives. We use it now for writing down the three canonical forms of second order partial differential equations

hyperbolic	$a_2 \phi_{xy} + b_2 \phi_x + c_2 \phi_y + f_2 \phi = g_2$
elliptic	$a_3 (\phi_{xx} + \phi_{yy}) + b_3 \phi_x + c_3 \phi_y + f_3 \phi = g_3$
parabolic	$a_4 \phi_{xx} + b_4 \phi_x + c_4 \phi_y + f_4 \phi = g_4.$

In these equations the a's, b's, c's, f's and g's are functions of the variables x, y. Laplace Transforms are useful in solving parabolic and some hyperbolic PDEs. They are not in general useful for solving elliptic PDEs.

Let us now turn to some practical examples in order to see how partial differential equations are solved. The commonest hyperbolic equation is the one dimensional wave equation. This takes the form

$$\frac{1}{c^2} \frac{\partial^2 u}{\partial t^2} = \frac{\partial^2 u}{\partial x^2}$$

where c is a constant called the *celerity* or *wave speed*. This equation can be used to describe waves travelling along a string, in which case u will represent the displacement of the string from equilibrium, x is distance along the string's equilibrium position, and t is time. As anyone who is familiar with string instruments will know, u takes the form of a wave. The derivation of this equation is not straightforward, but rests on the assumption that the displacement of the string from equilibrium is small. This means that x is virtually the distance along the string. If we think of the solution in (x, t) space, then the lines $x \pm ct =$ constant assume particular importance. They are called *characteristics* and the *general* solution to the wave equation has the general form

$$u = f(x - ct) + g(x + ct)$$

where f and g are arbitrary functions. If we expand f and g as Fourier series over the interval $[0, L]$ in which the string exists (for example between the bridge and the top (machine head) end of the fingerboard in a guitar) then it is immediate that u can be thought of as an infinite superposition of sinusoidal waves:-

$$u(x, t) = \sum_{n=0}^{\infty} a_n \cos[n(x - ct)] + b_n \sin[n(x - ct)]$$
$$+ \sum_{m=0}^{\infty} a'_m \cos[m(x + ct)] + b'_m \sin[m(x + ct)] + \frac{a_0}{\sqrt{2}} + \frac{a'_0}{\sqrt{2}}.$$

If the boundary conditions are appropriate to a musical instrument, i.e. $u = 0$ at $x = 0, L$ (all t) then this provides a good visual form of a Fourier series.

Although it is possible to use the Laplace Transform to solve such wave problems, this is rarely done as there are more natural methods and procedures that utilise the wave-like properties of the solutions but are outside the scope of this text. (What we are talking about here is the *method of characteristics* – see e.g. Williams (1980) Chapter 3.)

There is one particularly widely occurring elliptic partial differential equation which is mentioned here but cannot in general be solved using Laplace Transform techniques. This is Laplace's equation which, in its two dimensional form is

$$\frac{\partial^2 \phi}{\partial x^2} + \frac{\partial^2 \phi}{\partial y^2} = 0.$$

Functions ϕ that satisfy this equation are called *harmonic* and possess interesting mathematical properties. Perhaps the most important of these is the following. A function $\phi(x, y)$ which is harmonic in a domain $D \in \mathbf{R}^2$ has its maximum and minimum values on ∂D, the border of D, and not inside D itself. Laplace's equation occurs naturally in the fields of hydrodynamics, electromagnetic theory and elasticity when steady state problems are being solved in two dimensions. Examples include the analysis of standing water waves, the distribution of heat over a flat surface very far from the source a long time after the source has been switched on, and the vibrations of a membrane. Many of these problems are approximations to parabolic or wave problems that *can* be solved using Laplace Transforms. There are books devoted to the solutions of Laplace's equation, and the only reason its solution is mentioned here is because the properties associated with harmonic functions are useful in providing checks to solutions of parabolic or hyperbolic equations in some limiting cases. Let us without further ado go on to discuss parabolic equations.

The most widely occurring parabolic equation is called the heat conduction equation. In its simplest one dimensional form (using the two variables t (time) and x (distance)) it is written

$$\frac{\partial \phi}{\partial t} = \kappa \frac{\partial^2 \phi}{\partial x^2}.$$

This equation describes the manner in which heat $\phi(x, t)$ is conducted along a bar made of a homogeneous substance located along the x axis. The thermal conductivity (or thermal diffusivity) of the bar is a positive constant that has been labelled κ. One scenario is that the bar is cold (at room temperature say) and that heat has been applied to a point on the bar. The solution to this equation then describes the manner in which heat is subsequently distributed along the bar. Another possibility is that a severe form of heat, perhaps using a blowtorch, is applied to one point of the bar for a very short time then withdrawn. The solution of the heat conduction equation then shows how this heat gets conducted away from the site of the flame. A third possibility is that the rod is melting, and the equation is describing the way that the interface between the melted and unmelted rod is travelling away from the heat source that is causing the melting. Solving the heat conduction equation would predict the subsequent heat distribution, including the speed of travel of this interface. Each of these problems is what is called an *initial value problem* and this is precisely the kind of PDE that can be solved using Laplace Transforms. The bulk of the rest of this chapter is indeed devoted to solving these. One more piece of "housekeeping" is required however, that is the use of Fourier series in enabling boundary conditions to be satisfied. This in turn requires knowledge of the technique of separation of variables. This is probably revision, but in case it is not, the next section is devoted to it.

5.3 Separation of Variables

The technique of separating the variables will certainly be a familiar method for solving ordinary differential equations in the cases where the two variables can be isolated to occur exclusively on either side of the equality sign. In partial differential equations, the situation is slightly more complicated. It can be applied to Laplace's equation and to the wave equation, both of which were met in the last section. However, here we solve the heat conduction equation.

We consider the problem of solving the heat conduction equation together with the boundary conditions as specified below:

$$\frac{\partial \phi}{\partial t} = \kappa \frac{\partial^2 \phi}{\partial x^2}, \ x \in [0, L]$$
$$\phi(x, 0) = f(x) \text{ at time } t = 0$$
$$\phi(0, t) = \phi(L, t) = 0 \text{ for all time.}$$

The fundamental assumption for separating variables is to let

$$\phi(x, t) = T(t)X(x)$$

so that the heat conduction equation becomes

$$T'X = \kappa T X''$$

where prime denotes the derivative with respect to t or x. Dividing by XT we obtain

$$\frac{T'}{T} = \kappa \frac{X''}{X}.$$

The next step is crucial to understand. The left hand side is a function of t only, and the right hand side is a function of x only. *As t and x are independent variables, these must be equal to the same constant.* This constant is called the separation constant. It is wise to look ahead a little here. As the equation describes the very real situation of heat conduction, we should look for solutions that will decay as time progresses. This means that T is likely to decrease with time which in turn leads us to designate the separation constant as negative. Let it be $-\alpha^2$, so

$$\frac{T'}{T} = -\alpha^2 \text{ giving } T(t) = T_0 e^{-\alpha^2 t}, \quad t \geq 0,$$

and

$$\frac{X''}{X} = -\frac{\alpha^2}{\kappa} \text{ giving } X(x) = a' \cos\left(\frac{\alpha x}{\sqrt{\kappa}}\right) + b' \sin\left(\frac{\alpha x}{\sqrt{\kappa}}\right).$$

Whence the solution is

$$\phi(x, t) = e^{-\alpha^2 t} \left(a \cos\left(\frac{\alpha x}{\sqrt{\kappa}}\right) + b \sin\left(\frac{\alpha x}{\sqrt{\kappa}}\right) \right).$$

At time $t = 0$, $\phi(x, 0) = f(x)$ which is some prescribed function of x (the initial temperature distribution along a bar $x \in [0, L]$ perhaps) then we would seem to require that

$$f(x) = a \cos\left(\frac{\alpha x}{\sqrt{\kappa}}\right) + b \sin\left(\frac{\alpha x}{\sqrt{\kappa}}\right)$$

which in general is not possible. However, we can now use the separation constant to our advantage. Recall that in Chapter 4 it was possible for any piecewise continuous function $f(x), x \in [0, L]$ to be expressed as a series of trigonometric functions. In particular we can express $f(x)$ by

$$f(x) \sim \sum_{n=1}^{\infty} b_n \sin\left(\frac{\alpha_n x}{\sqrt{\kappa}}\right),$$

writing $\alpha = \alpha_n$ to emphasise its n dependence. Further, if we set

$$\frac{\alpha_n}{\sqrt{\kappa}} = \frac{n\pi}{L}, \quad n = \text{ integer}$$

then the boundary conditions at $x = 0$ and $x = L$ are both satisfied. Here we have expressed the function $f(x)$ as a half range Fourier sine series which is consistent with the given boundary conditions. Half range cosine series or full range series can also be used of course depending on the problem.

Here, this leads to the complete solution of this particular problem in terms of the series

$$\phi(x, t) = \sum_{n=1}^{\infty} b_n e^{-(n^2\pi^2\kappa/L^2)t} \sin\left(\frac{n\pi x}{L}\right),$$

where

$$b_n = \frac{2}{L} \int_0^L f(x) \sin\left(\frac{n\pi x}{L}\right) dx.$$

The solution can be justified as correct using the following arguments. First of all the heat conduction equation is linear which means that we can superpose the separable solutions in the form of a series to obtain another solution provided the series is convergent. This convergence is established easily since the Fourier series itself is pointwise convergent by construction, and the multiplying factor $e^{-(n^2\pi^2\kappa/L^2)t}$ is always less than or equal to one. The $\phi(x, t)$ obtained is thus the solution to the heat conduction equation. Here is a specific example.

Example 5.2 *Determine the solution to the boundary value problem:*

$$\frac{\partial \phi}{\partial t} = \kappa \frac{\partial^2 \phi}{\partial x^2}, \quad x \in [0, 2]$$
$$\phi(x, 0) = x + (2 - 2x)H(x - 1) \quad \text{at time} \quad t = 0$$
$$\phi(0, t) = \phi(2, t) = 0 \quad \text{for all time}$$

where $H(x)$ is Heaviside's Unit Step Function (see Chapter 2, page 19).

Solution The form of the function $\phi(x, 0)$ is displayed in Figure 5.1. This function is expressed as a Fourier sine series by the methods outlined in Chapter 4. This is in order to make automatic the satisfying of the boundary conditions. The Fourier sine series is not derived in detail as this belongs in Chapter 4. The result is

$$x + (2 - 2x)H(x - 1) \sim \sum_{n=1}^{\infty} \frac{8(-1)^{n-1}}{\pi^2(2n - 1)^2} \sin\left(\frac{n\pi x}{2}\right).$$

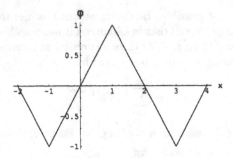

Figure 5.1: The initial distribution of $\phi(x,t)$ drawn as an odd function

The solution to this particular boundary value problem is therefore

$$\phi(x,t) = \sum_{n=1}^{\infty} \frac{8(-1)^{n-1}}{\pi^2 (2n-1)^2} e^{-(n^2\pi^2\kappa/L^2)t} \sin\left(\frac{n\pi x}{2}\right).$$

There is much more on solving PDEs using separation of variables in specialist texts such as Zauderer (1989). We shall not dwell further on the method for its own sake, but move on to solve PDEs using Laplace Transforms which itself often requires use of separation of variables.

5.4   Using Laplace Transforms to Solve PDEs

We now turn to solving the heat conduction equation using Laplace Transforms. The symbol t invariably denotes time in this equation and the fact that time runs from now ($t = 0$) to infinity neatly coincides with the range of the Laplace Transform. Remembering that

$$\mathcal{L}\{\phi'\} = \int_0^\infty e^{-st} \frac{d\phi}{dt} dt = s\bar{\phi}(s) - \phi(0)$$

where overbar (as usual) denotes the Laplace Transform (i.e. $\mathcal{L}(\phi) = \bar{\phi}$) gives us the means of turning the PDE

$$\frac{\partial \phi}{\partial t} = \kappa \frac{\partial^2 \phi}{\partial x^2}$$

into the ODE

$$s\bar{\phi}(s) - \phi(0) = \kappa \frac{d^2 \bar{\phi}}{dx^2}.$$

We have of course assumed that

$$\mathcal{L}\left\{\kappa \frac{\partial^2 \phi}{\partial x^2}\right\} = \int_0^\infty \kappa \frac{\partial^2 \phi}{\partial x^2} e^{-st} dt = \kappa \frac{d^2}{dx^2} \int_0^\infty \phi e^{-st} dt = \kappa \frac{d^2 \bar{\phi}}{dx^2}$$

which demands a continuous second order partial derivative with respect to x and an improper integral with respect to t which are well defined for all values of x so that the legality of interchanging differentiation and integration (sometimes called Leibniz' Rule) is ensured. Rather than continue in general terms, having seen how the Laplace Transform can be applied to a PDE let us do an example.

Example 5.3 *Solve the heat conduction equation*

$$\frac{\partial \phi}{\partial t} = \frac{\partial^2 \phi}{\partial x^2}$$

in the region $t > 0$, $x > 0$ with boundary conditions $\phi(x, 0) = 0$ $x > 0$ (initial condition), $\phi(0, t) = 1$, $t > 0$ (temperature held constant at the origin) and $\lim_{x \to \infty} \phi(x, t) = 0$ (the temperature a long way from the origin remains at its initial value). [For an alternative physical interpretation of this, see the end of the solution.]

Solution Taking the Laplace Transform (in t of course) gives

$$s\bar{\phi} - \phi(0) = \frac{d^2 \bar{\phi}}{dx^2}$$

or

$$s\bar{\phi} = \frac{d^2 \bar{\phi}}{dx^2}$$

since $\phi(x, 0) = 0$. This is an ODE with constant coefficients (not dependent on x; the presence of s does not affect this argument as there are no s derivatives) with solution

$$\bar{\phi}(x, s) = Ae^{-x\sqrt{s}} + Be^{x\sqrt{s}}.$$

Now, since $\lim_{x \to \infty} \phi(x, t) = 0$ and

$$\bar{\phi}(x, s) = \int_0^\infty \phi(x, t)e^{-st}dt$$

we have that $\bar{\phi}(x, s) \to 0$ as $x \to \infty$, hence $B = 0$. Thus

$$\bar{\phi}(x, s) = Ae^{-x\sqrt{s}}.$$

Letting $x = 0$ in the Laplace Transform of ϕ gives

$$\bar{\phi}(0, s) = \int_0^\infty \phi(0, t)e^{-st}dt = \int_0^\infty e^{-st}dt = \frac{1}{s}$$

given that $\phi(0, t) = 1$ for all t. Notice that the "for all t" is crucial as the Laplace Transform has integrated through all positive values of t. Inserting this boundary condition for $\bar{\phi}$ gives

$$A = \frac{1}{s}$$

whence the solution for $\bar{\phi}$ is

$$\bar{\phi}(x,s) = \frac{e^{-x\sqrt{s}}}{s}.$$

Inverting this using a table of standard forms, or Example 3.11, gives

$$\phi(x,t) = \text{erfc}\left(\frac{1}{2}xt^{-1/2}\right).$$

Another physical interpretation of this solution runs as follows. Suppose there is a viscous fluid occupying the region $x > 0$ with a flat plate on the plane $x = 0$. (Think of the plate as vertical to eliminate the effects of gravity.) The above expression for ϕ is the solution to jerking the plate in its own plane so that a velocity near the plate is generated. Viscosity takes on the role of conductivity here, the value is taken as unity in the above example. ϕ is the fluid speed dependent on time t and vertical distance from the plate x.

Most of you will realise at once that using Laplace Transforms to solve this kind of PDE is not a problem apart perhaps from evaluating the Inverse Laplace Transform. There is a general formula for the Inverse Laplace Transform and we meet this informally in the next chapter and more formally in Chapter 7. You will also see that the solution

$$\phi(x,t) = \text{erfc}\left(\frac{1}{2}xt^{-1/2}\right)$$

is in no way expressible in variable separable form. This particular problem is not amenable to solution using separation of variables because of the particular boundary conditions. There is a method of solution whereby we set $\xi = xt^{-1/2}$ and transform the equation $\phi_t = \phi_{xx}$ from (x,t) space into ξ space. This is called using a similarity variable. Further details can be found in more specialised texts, for example Zauderer (1989).

One more subject that is essential to consider theoretically but not discussed here is uniqueness. The heat conduction equation has a unique solution provided boundary conditions are given at $t = 0$ (initial condition) together with values of ϕ (or its derivative with respect to x) at $x = 0$ and another value ($x = a$ say) for all t. The proof of this is straightforward and makes use of contradiction. This is quite typical of uniqueness proofs, and the details can be found in, for example, Williams (1980) pp195–199. Although the proof of uniqueness is a side issue for a book on Laplace Transforms and Fourier series, it is always important. Once a solution to a boundary value problem like the heat conduction in a bar has been found, it is vital to be sure that it is the only one. Problems that do not have a unique solution are called *ill posed* and they are not dealt with in this text. In the last twenty years, a great deal of attention has been focused on non-linear problems. These do not have a unique solution (in general) but are as far as we know accurate descriptions of real problems.

So far in this chapter we have discussed the solution of evolutionary partial differential equations, typically the heat conduction equation where initial conditions dictate the behaviour of the solution. Laplace Transforms can also be used to solve second order hyperbolic equations typified by the wave equation. Let us see how this is done. The one-dimensional wave equation may be written

$$\frac{\partial^2 \phi}{\partial t^2} = c^2 \frac{\partial^2 \phi}{\partial x^2}$$

where c is a constant called the celerity or wave speed, x is the displacement and t of course is time. The kind of problem that can be solved is that for which conditions at time $t = 0$ are specified. This does not cover all wave problems and may not even be considered to be typical of wave problems, but although problems for which conditions at two different times are given can be solved using Laplace Transforms (see Chapter 3) alternative solution methods (e.g. using characteristics) are usually better. As before, the technique is to take the Laplace Transform of the equation with respect to time, noting that this time there is a second derivative to deal with. Defining

$$\bar{\phi} = \mathcal{L}\{\phi\}$$

and using

$$\mathcal{L}\left\{\frac{\partial^2 \phi}{\partial t^2}\right\} = s^2 \bar{\phi} - s\phi(0) - \phi'(0)$$

the wave equation transforms into the following ordinary differential equation for $\bar{\phi}$:

$$s^2 \bar{\phi} - s\phi(0) - \phi'(0) = c^2 \frac{d^2 \bar{\phi}}{dx^2}.$$

The solution to this equation is conveniently written in terms of hyperbolic functions since the x boundary conditions are almost always prescribed at two finite values of x. If the problem is in an infinite half space, for example the vibration of a very long beam fixed at one end (a cantilever), the complementary function $Ae^{sx/c} + Be^{-sx/c}$ can be useful. As with the heat conduction equation, rather than pursuing the general problem, let us solve a specific example.

Example 5.4 *A beam of length a initially at rest has one end fixed at $x = 0$ with the other end free to move at $x = a$. Assuming that the beam only moves longitudinally (in the x direction) and is subject to a constant force ED along its length where E is the Young's modulus of the beam and D is the displacement per unit length, find the longitudinal displacement of the beam at any subsequent time. Find also the motion of the free end at $x = a$.*

Solution The first part of this problem has been done already in that the Laplace Transform of the one dimensional wave equation is

$$s^2 \bar{\phi} - s\phi(0) - \phi'(0) = c^2 \frac{d^2 \bar{\phi}}{dx^2}$$

where $\phi(x, t)$ is now the displacement, $\bar{\phi}$ its Laplace Transform and we have assumed that the beam is perfectly elastic, hence actually obeying the one dimensional wave equation. The additional information we have available is that the beam is initially at rest, $x = 0$ is fixed for all time and that the other end at $x = a$ is free. These are translated into mathematics as follows:-

$$\phi(x,0) = 0, \quad \frac{\partial \phi}{\partial t}(x,0) = 0 \text{ (beam is initially at rest)}$$

together with

$$\phi(0,t) = 0 \ (x = 0 \text{ fixed for all time})$$

and

$$\frac{\partial \phi(a,t)}{\partial x} = D \ \forall t \text{ (the end at } x = a \text{ is free).}$$

Hence

$$\frac{d^2 \bar{\phi}}{dx^2} = \frac{s^2}{c^2} \bar{\phi}$$

the solution of which can be written

$$\bar{\phi}(x,s) = k_1 \cosh\left(\frac{sx}{c}\right) + k_2 \sinh\left(\frac{sx}{c}\right).$$

In order to find k_1 and k_2, we use the x boundary condition. If $\phi(0,t) = 0$ then, taking the Laplace Transform, $\bar{\phi}(0,s) = 0$ so $k_1 = 0$. Hence

$$\bar{\phi}(x,s) = k_2 \sinh\left(\frac{sx}{c}\right) \text{ and } \frac{d\bar{\phi}}{dx} = \frac{sk_2}{c} \cosh\left(\frac{sx}{c}\right).$$

We have that

$$\frac{\partial \phi}{\partial x} = D \text{ at } x = a \text{ for all } t.$$

The Laplace Transform of this is

$$\frac{d\bar{\phi}}{dx} = \frac{D}{s} \text{ at } x = a.$$

Hence

$$\frac{D}{s} = \frac{sk_2}{c} \cosh\left(\frac{sa}{c}\right)$$

from which

$$k_2 = \frac{Dc}{s^2 \cosh(\frac{sa}{c})}.$$

Hence the solution to the ODE for $\bar{\phi}$ is

$$\bar{\phi} = \frac{Dc \sinh\left(\frac{sx}{c}\right)}{s^2 \cosh\left(\frac{sa}{c}\right)}$$

and is completely determined.

The remaining problem, and it is a significant one, is that of inverting this Laplace Transform to obtain the solution $\phi(x,t)$. Many would use the inversion

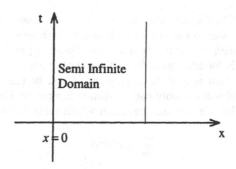

Figure 5.2: The domain

formula of Chapter 7, but as this is not available to us, we scan the table of standard Laplace Transforms and find the result

$$\mathcal{L}^{-1}\left\{\frac{\sinh(sx)}{s^2\cosh(sa)}\right\} = x + \frac{8a}{\pi^2}\sum_{n=1}^{\infty}\frac{(-1)^n}{(2n-1)^2}\sin\left(\frac{(2n-1)\pi x}{2a}\right)\cos\left(\frac{(2n-1)\pi t}{2a}\right)$$

and deduce that

$$\phi(x,t) = Dx + \frac{8aD}{\pi^2}\sum_{n=1}^{\infty}\frac{(-1)^n}{(2n-1)^2}\sin\left(\frac{(2n-1)\pi x}{2a}\right)\cos\left(\frac{(2n-1)\pi ct}{2a}\right).$$

It is not until we meet the general inversion formula in Chapter 7 that we are able to see how such formulae are derived using complex analysis.

5.5 Boundary Conditions and Asymptotics

A partial differential equation together with enough boundary conditions to ensure the existence of a unique solution is called a boundary value problem, sometimes abbreviated to BVP. Parabolic and hyperbolic equations of the type suited to solution using Laplace Transforms are defined over a semi infinite domain. In two dimensional Cartesian co-ordinates, this domain will take the form of a semi infinite strip shown in Figure 5.2. Since one of the variables is almost always time, the problem has conditions given at time $t = 0$ and the solution once found is valid for all subsequent times, theoretically to time $t = \infty$. Indeed, this is the main reason why Laplace Transforms are so useful for these types of problems. Thus, in transformed space, the boundary condition at $t = 0$ and the behaviour of the solution as $t \to \infty$ get swept up in the Laplace Transform and appear explicitly in the transformed equations, whereas the conditions on the other variable (x say) which form the two infinite sides of the rectangular domain of Figure 5.2 get transformed into two boundary conditions that are the end points of a two point boundary value problem (one dimensional) in x. The Laplace Transform variable s becomes passive because no derivatives with respect to s are present. If there are three variables (or perhaps more) and one of them is time-like in that conditions are prescribed at $t = 0$ and the domain extends for arbitrarily large time, then similar arguments prevail. However, the situation is a little more complicated in that the transformed boundary value problem has only had its dimension reduced by one. That is, a three dimensional heat conduction equation which takes the form

$$\frac{\partial \phi}{\partial t} = \kappa \nabla^2 \phi$$

where

$$\nabla^2 \phi = \frac{\partial^2 \phi}{\partial x^2} + \frac{\partial^2 \phi}{\partial y^2} + \frac{\partial^2 \phi}{\partial z^2}$$

which is in effect a four dimensional equation involving time together with three space dimensions, is transformed into a Poisson like equation:-

$$\kappa \nabla^2 \bar{\phi}(x, y, z, s) = s\bar{\phi}(x, y, z, s) - \phi(x, y, z, 0)$$

where, as usual, $\mathcal{L}\{\phi(x, y, z, t)\} = \bar{\phi}(x, y, z, s)$.

It remains difficult in general to determine the Inverse Laplace Transform, so various properties of the Laplace Transform are invoked as an alternative to complete inversion. One device is particularly useful and amounts to using a special case of Watson's Lemma, a well known result in asymptotic analysis. If for small values of t, $\phi(x, y, z, t)$ has a power series expansion of the form

$$\phi(x, y, z, t) = \sum_{n=0}^{\infty} a_n(x, y, z) t^{k+n}$$

and $|\phi|e^{-ct}$ is bounded for some k and c, then the result of Chapter 2 following Theorem 2.13 can be invoked and we can deduce that as $s \to \infty$ the Laplace Transform of ϕ, $\bar{\phi}$ has an equivalent asymptotic expansion

$$\bar{\phi}(x, y, z, s) = \sum_{n=0}^{\infty} a_n(x, y, z) \frac{\Gamma(n + k + 1)}{s^{n+k+1}}.$$

Now asymptotic expansions are not necessarily convergent series; however the first few terms do give a good approximation to the function $\bar{\phi}(x, y, z, s)$. What we have here is an approximation to the transform of ϕ that relates the form for large s to the behaviour of the original variable ϕ for small t. Note that this is consistent with the initial value theorem (Theorem 2.12). It is sometimes the case that each series is absolutely convergent. A term by term evaluation can then be justified. However, often the most interesting and useful applications of asymptotic analysis take place when the series are not convergent. The classical text by Copson (1967) remains definitive. The serious use of these results demands a working knowledge of complex variable theory, in particular of poles of complex functions and residue theory. These are not dealt with until

Chapter 7 so examples involving complex variables are postponed until then. Here is a reasonably straightforward example using asymptotic series, just to get the idea.

Example 5.5 *Find an approximation to the solution of the partial differential equation*

$$\frac{\partial \phi}{\partial t} = c^2 \frac{\partial^2 \phi}{\partial x^2}$$

for small times where $\phi(x,0) = \cos(x)$, *by using an asymptotic series.*

Solution It is possible to solve this BVP exactly, but let us take Laplace Transforms to obtain

$$s\bar{\phi} - \cos(x) = c^2 \frac{d^2 \bar{\phi}}{dx^2}$$

then try an asymptotic series of the form

$$\bar{\phi}(x,s) = \sum_{n=0}^{\infty} \frac{b_n(x)}{s^{n+k+1}}$$

valid far enough away from the singularities of $\bar{\phi}$ in s space. (This is only true provided $\bar{\phi}$ does not have branch points. See Section 7.4 for more about branch points.) Equating coefficients of $1/s^n$ yields straight away that $k = 0$, then we get

$$b_0 = \cos(x); \quad b_1 = c^2 \frac{d^2 b_0}{dx^2}; \quad b_2 = c^2 \frac{d^2 b_1}{dx^2}; \ldots$$

and hence

$$b_1 = -c^2 \cos(x), \quad b_2 = c^4 \cos(x), \quad b_3 = -c^6 \cos(x) \text{ etc.}$$

This yields

$$\bar{\phi}(x,s) = \cos(x) \sum_{n=0}^{\infty} \frac{(-1)^n c^{2n}}{s^{n+2}}.$$

Provided we have $s > c > 0$, term by term inversion is allowable here as the series will then converge for all values of x. It is uniformly but not absolutely convergent. This results in

$$\phi(x,t) = \cos(x) \sum_{n=0}^{\infty} \frac{(-1)^n c^{2n} t^n}{n!}$$

which is immediately recognised as

$$\phi(x,t) = \cos(x) e^{-c^2 t}$$

a solution that could have been obtained directly using separation of variables. As is obvious from this last example, we are hampered in the kind of problem that can be solved because we have yet to gain experience of the use of complex variables. Fourier Transforms are also a handy tool for solving certain types of BVP, and these are the subject of the next chapter. So finally in this chapter, a word about other methods of solving partial differential equations.

In the years since the development of the workstation and desk top micro-computer, there has been a revolution in the methods used by engineers and applied scientists in industry who need to solve boundary value problems. In real situations, these problems are governed by equations that are far more complicated than those covered here. Analytical methods can only give a first approximation to the solution of such problems and these methods have been surpassed by numerical methods based on finite difference and finite element approximations to the partial differential equations. However, it is still essential to retain knowledge of analytical methods, as these give insight as to the general behaviour in a way that a numerical solution will never do, and because analytical methods actually can lead to an increase in efficiency in the numerical method eventually employed to solve the real problem. For example, an analytical method may be able to tell where a solution changes rapidly even though the solution itself cannot be obtained in closed form and this helps to design the numerical procedure, perhaps even suggesting alternative methods (finite elements instead of finite differences) but more likely helping to decide on co-ordinate systems and step lengths. The role of mathematics has thus changed in emphasis from providing direct solutions to real problems to giving insight into the underlying properties of the solutions. As far as this chapter is concerned, future applications of Laplace Transforms to partial differential equations are met briefly in Chapter 7, but much of the applications are too advanced for this book and belong in specialist texts on the subject, e.g. Weinberger (1965).

5.6   Exercises

1. Using separation of variables, solve the boundary value problem:

$$\frac{\partial \phi}{\partial t} = \kappa \frac{\partial^2 \phi}{\partial x^2}, \quad x \in \left[0, \frac{\pi}{4}\right]$$

$$\phi(x,0) = x\left(\frac{\pi}{4} - x\right) \quad \text{at time} \quad t = 0$$

$$\phi(0,t) = \phi\left(\frac{\pi}{4}, t\right) = 0 \quad \text{for all time,}$$

using the methods of Chapter 4 to determine the Fourier series representation of the function $x\left(\frac{\pi}{4} - x\right)$.

2. The function $\phi(x,t)$ satisfies the PDE

$$a\frac{\partial^2 \phi}{\partial x^2} - b\frac{\partial \phi}{\partial x} - \frac{\partial \phi}{\partial t} = 0$$

with $x > 0$, $a > 0$, $b > 0$ and boundary conditions $\phi(x,0) = 0$ for all x, $\phi(0,t) = 1$ for all t and $\phi \to 0$ as $x \to \infty$, $t > 0$. Use Laplace Transforms to find $\bar{\phi}$, the Laplace Transform of ϕ. (Do not attempt to invert it.)

3. Use Laplace Transforms to solve again the BVP of Exercise 1 but this time in the form

$$\phi(x,t) = -x^2 + \frac{\pi}{4}x - 2\kappa t + 2\kappa \mathcal{L}^{-1} \left\{ \frac{\sinh(\frac{\pi}{4} - x)\sqrt{\frac{s}{\kappa}}}{s^2 \sinh(\frac{\pi}{4})\sqrt{\frac{s}{\kappa}}} \right\}$$

$$+ 2\kappa \mathcal{L}^{-1} \left\{ \frac{\sinh(x\sqrt{\frac{s}{\kappa}})}{s^2 \sinh(\frac{\pi}{4}\sqrt{\frac{s}{\kappa}})} \right\}.$$

Use the table of Laplace Transforms to invert this expression. Explain any differences between this solution and the answer to Exercise 1.

4. Solve the PDE

$$\frac{\partial^2 \phi}{\partial x^2} = \frac{\partial \phi}{\partial y}$$

with boundary conditions $\phi(x,0) = 0$, $\phi(0,y) = 1$, $y > 0$ and

$$\lim_{x \to \infty} \phi(x,y) = 0.$$

5. Suppose that $u(x,t)$ satisfies the equation of telegraphy

$$\frac{1}{c^2} \frac{\partial^2 u}{\partial t^2} - \frac{k}{c^2} \frac{\partial u}{\partial t} + \frac{1}{4} \frac{k^2}{c^2} u = \frac{\partial^2 u}{\partial x^2}.$$

Find the equation satisfied by $\phi = ue^{-kt/2}$, and hence use Laplace Transforms (in t) to determine the solution for which

$$u(x,0) = \cos(mx), \quad \frac{\partial u}{\partial t}(x,0) = 0 \text{ and } u(0,t) = e^{kt/2}.$$

6. The function $u(x,t)$ satisfies the BVP

$$u_t - c^2 u_{xx} = 0, \quad x > 0, \quad t > 0, \quad u(0,t) = f(t), \quad u(x,0) = 0$$

where $f(t)$ is piecewise continuous and of exponential order. (The suffix derivative notation has been used.) Find the solution of this BVP by using Laplace Transforms together with the convolution theorem. Determine the explicit solution in the special case where $f(t) = \delta(t)$, where $\delta(t)$ is Dirac's delta function.

7. A semi-infinite solid occupying the region $x > 0$ has its initial temperature set to zero. A constant heat flux is applied to the face at $x = 0$, so that $T_x(0,t) = -\alpha$ where T is the temperature field and α is a constant. Assuming linear heat conduction, find the temperature at any point x ($x > 0$) of the bar and show that the temperature at the face at time t is given by

$$\alpha \sqrt{\frac{\kappa}{\pi t}}$$

where κ is the thermal conductivity of the bar.

8. Use asymptotic series to provide an approximate solution to the wave equation

$$\frac{\partial^2 u}{\partial t^2} = c^2 \frac{\partial^2 u}{\partial x^2}$$

valid for small values of t with

$$u(x,0) = 0, \quad \frac{\partial u}{\partial t}(x,0) = \cos(x).$$

9. Repeat the last exercise, but using instead the boundary conditions

$$u(x,0) = \cos(x), \quad \frac{\partial u}{\partial t}(x,0) = 0.$$

6
Fourier Transforms

6.1   Introduction

Later in this chapter we define the Fourier Transform. There are two ways of approaching the subject of Fourier Transforms, both ways are open to us! One way is to carry on directly from Chapter 4 and define Fourier Transforms in terms of the mathematics of linear spaces by carefully increasing the period of the function $f(x)$. This would lead to the Fourier series we defined in Chapter 4 becoming, in the limit of infinite period, an integral. This integral leads directly to the Fourier Transform. On the other hand, the Fourier Transform can be straightforwardly defined as an example of an integral transform and its properties compared and in many cases contrasted with those of the Laplace Transform. It is this second approach that is favoured here, with the first more pure mathematical approach outlined towards the end of Section 6.2. This choice is arbitrary, but it is felt that the more "hands on" approach should dominate here. Having said this, texts that concentrate on computational aspects such as the FFT (Fast Fourier Transform), on time series analysis and on other branches of applied statistics sometimes do prefer the more pure approach in order to emphasise precision.

6.2   Deriving the Fourier Transform

Definition 6.1 *Let f be a function defined for all $x \in \mathbf{R}$ with values in $\mathbf{C}$. The Fourier Transform is a mapping $F : \mathbf{R} \to \mathbf{C}$ defined by*

$$F(\omega) = \int_{-\infty}^{\infty} f(x) e^{-i\omega x} dx.$$

Of course, for some $f(x)$ the integral on the right does not exist. We shall spend some time discussing this a little later. There can be what amounts to trivial differences between definitions involving factors of 2π or $\sqrt{2\pi}$. Although this is of little consequence mathematically, it is important to stick to the definition whichever version is chosen. In engineering where x is often time, and ω frequency, factors of 2π or $\sqrt{2\pi}$ can make a lot of difference.

If $F(\omega)$ is defined by the integral above, then it can be shown that

$$f(x) = \frac{1}{2\pi} \int_{-\infty}^{\infty} F(\omega)e^{i\omega x} d\omega.$$

This is the inverse Fourier Transform. It is instructive to consider $F(\omega)$ as a complex valued function of the form

$$F(\omega) = A(\omega)e^{i\phi(\omega)}$$

where $A(\omega)$ and $\phi(\omega)$ are real functions of the real variable ω. F is thus a complex valued function of a real variable ω. Some readers will recognise $F(\omega)$ as a spectrum function, hence the letters A and ϕ which represent the amplitude and phase of F respectively. We shall not dwell on this here however. If we merely substitute for $F(\omega)$ we obtain

$$f(x) = \frac{1}{2\pi} \int_{-\infty}^{\infty} A(\omega)e^{i\omega x + \phi(\omega)} d\omega.$$

We shall return to this later when discussing the relationship between Fourier Transforms and Fourier series. Let us now consider what functions permit Fourier Transforms. A glance at the definition tells us that we cannot for example calculate the Fourier Transform of polynomials or even constants due to the oscillatory nature of the kernel. This is a feature that might seem to render the Fourier Transform useless. It is certainly a difficulty, but one that is more or less completely solved by extending what is meant by an integrable function through the use of generalised functions. These were introduced in Section 2.6, and it turns out that the Fourier Transform of a constant is closely related to the Dirac-δ function defined in Section 2.6. The impulse function is a representative of this class of functions and we met many of its properties in Chapter 2. In that chapter, mention was also made of the use of the impulse function in many applications, especially in electrical engineering and signal processing. The general mathematics of generalised functions is outside the scope of this text, but more of its properties will be met later in this chapter.

If we write the function to be transformed in the form $e^{-kx}f(x)$ then the Fourier Transform is the integral

$$\int_{-\infty}^{\infty} e^{-i\omega x} e^{-kx} f(x) dx$$

straight from the definition. In this form, the Fourier Transform can be related to the Laplace Transform. First of all, write

$$F_k(\omega) = \int_{0}^{\infty} e^{-(k+i\omega)x} f(x) dx$$

then $F_k(\omega)$ will exist provided the function $f(x)$ is of exponential order (see Chapter 1). Note too that the bottom limit has become 0. This reflects that the variable x is usually time. The inverse of $F_k(\omega)$ is straightforward to find once it is realised that the function $f(x)$ can be defined as identically zero for $x < 0$. Whence we have

$$\frac{1}{2\pi} \int_{-\infty}^{\infty} e^{i\omega x} F_k(\omega) d\omega = \begin{cases} 0 & x < 0 \\ e^{-kx} f(x) & x \geq 0. \end{cases}$$

An alternative way of expressing this inverse is in the form

$$\frac{1}{2\pi} \int_{-\infty}^{\infty} e^{(k+i\omega)x} F_k(\omega) d\omega = \begin{cases} 0 & x < 0 \\ f(x) & x \geq 0. \end{cases}$$

In this formula, and in the one above for $F_k(\omega)$, the complex number $k + i\omega$ occurs naturally. This is a variable, and therefore a *complex* variable and it corresponds to s the Laplace Transform variable defined in Chapter 1. Now, the integral on the left of the last equation is not meaningful if we are to regard $k + i\omega = s$ as the variable. As k is not varying, we can simply write

$$ds = id\omega$$

and s will take the values $k - i\infty$ and $k + i\infty$ at the limits $\omega = -\infty$ and $\omega = \infty$ respectively. The left-hand integral is now, when written as an integral in s,

$$\frac{1}{2\pi i} \int_{k-i\infty}^{k+i\infty} e^{sx} F(s) ds$$

where $F(s) = F_k(\omega)$ is now a complex valued function of the complex variable s. Although there is nothing illegal in the way the variable changes have been undertaken in the above manipulations, it does amount to a rather cartoon derivation. A more rigorous derivation of this integral is given in the next chapter after complex variables have been properly introduced. The formula

$$f(x) = \frac{1}{2\pi i} \int_{k-i\infty}^{k+i\infty} e^{sx} F(s) ds$$

is indeed the general form of the inverse Laplace Transform, given $F(s) = \mathcal{L}\{f(x)\}$.

We now approach the definition of Fourier Transforms from a different viewpoint. In Chapter 4, Fourier series were discussed at some length. As a summary for present purposes, if $f(x)$ is a periodic function, and for simplicity let us take the period as being 2π (otherwise in all that follows replace x by $lx/2\pi$ where l is the period) then $f(x)$ can be expressed as the Fourier series

$$f(x) \sim \frac{a_0}{2} + \sum_{n=1}^{\infty} (a_n \cos(nx) + b_n \sin(nx))$$

where

$$a_n = \frac{1}{\pi} \int_{-\pi}^{\pi} f(x)\cos(nx)dx, \quad n = 0, 1, 2, \ldots$$

and

$$b_n = \frac{1}{\pi} \int_{-\pi}^{\pi} f(x)\sin(nx)dx \quad n = 0, 1, 2, \ldots.$$

These have been derived in Chapter 4 and follow from the orthogonality properties of sine and cosine. The factor $\frac{1}{2}$ in the constant term enables a_n to be expressed as the integral shown without $n = 0$ being an exception. It is merely a convenience, and as with factors of 2π in the definition of Fourier Transform, the definition of a Fourier series can have these trivial differences. The task now is to see how Fourier series can be used to generate not only the Fourier Transform but also its inverse. The first step is to convert sine and cosine into exponential form; this will re-derive the complex form of the Fourier series first done in Chapter 4. Such a re-derivation is necessary because of the slight change in definition of Fourier series involving the a_0 term. So we start with the standard formulae

$$\cos(nx) = \frac{1}{2}(e^{inx} + e^{-inx})$$

and

$$\sin(nx) = \frac{1}{2i}(e^{inx} - e^{-inx}).$$

Some algebra of an elementary nature is required before the Fourier series as given above is converted into the complex form

$$f(x) = \sum_{n=-\infty}^{\infty} c_n e^{inx}$$

where

$$c_n = \frac{1}{2\pi} \int_{-\pi}^{\pi} f(x)e^{-inx}dx.$$

The complex numbers c_n are related to the real numbers a_n and b_n by the simple formulae

$$c_n = \frac{1}{2}(a_n - ib_n), \quad c_{-n} = \overline{c_n}, \quad n = 0, 1, 2, \ldots$$

and it is assumed that $b_0 = 0$. The overbar denotes complex conjugate. There are several methods that enable one to move from this statement of complex Fourier series to Fourier Transforms. The method adopted here is hopefully easy to follow as it is essentially visual. First of all consider a function $g(t)$ which has period T. (We have converted from x to t as the period is T and no longer 2π. It is the transformation $x \to 2\pi t/T$.) Figure 6.1 gives some insight into what we are trying to do here. The functions $f(t)$ and $g(t)$ coincide precisely

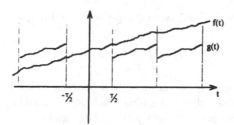

Figure 6.1: The function $f(t)$ and its periodic clone $g(t)$.

in the interval $[-\frac{T}{2}, \frac{T}{2}]$, but not necessarily outside this range. Algebraically we can write

$$g(t) = \begin{cases} f(t) & |t| < \frac{1}{2}T \\ f(t - nT) & \frac{1}{2}(2n - 1)T < |t| < \frac{1}{2}(2n + 1)T \end{cases}$$

where n is an integer. Since $g(t)$ is periodic, it possesses a Fourier series which, using the complex form just derived, can be written

$$g(t) = \sum_{n=-\infty}^{\infty} G_n e^{in\omega_0 t}$$

where G_n is given by

$$G_n = \frac{1}{T} \int_{-T/2}^{T/2} g(t) e^{-in\omega_0 t} dt$$

and $\omega_0 = 2\pi/T$ is the frequency. Again, this is obtained straightforwardly from the previous results in x by writing

$$x = \frac{2\pi t}{T} = \omega_0 t.$$

We now combine these results to give

$$g(t) = \sum_{n=-\infty}^{\infty} \left[\frac{1}{T} \int_{-T/2}^{T/2} g(t) e^{-in\omega_0 t} dt \right] e^{in\omega_0 t}.$$

The next step is the important one. Note that

$$n\omega_0 = \frac{2\pi n}{T}$$

and that the difference in frequency between successive terms is ω_0. As we need this to get smaller and smaller, let $\omega_0 = \Delta\omega$ and $n\omega_0 = \omega_n$ which is to remain finite as $n \to \infty$ and $\omega_0 \to 0$ together. The integral for G_n can thus be re-written

$$G = \int_{-T/2}^{T/2} g(t)e^{-i\omega_n t}dt.$$

Having set everything up, we are now in a position to let $T \to \infty$, the mathematical equivalent of lighting the blue touchpaper! Looking at Figure 6.1 this means that the functions $f(t)$ and $g(t)$ coincide, and

$$g(t) = \lim_{T \to \infty} \sum_{n=-\infty}^{\infty} Ge^{i\omega_n t}\frac{\Delta\omega}{2\pi}$$

$$= \frac{1}{2\pi}\int_{-\infty}^{\infty} Ge^{i\omega t}d\omega$$

with

$$G(\omega) = \int_{-\infty}^{\infty} g(t)e^{-i\omega t}dt.$$

We have let $T \to \infty$, replaced $\Delta\omega$ by the differential $d\omega$ and ω_n by the variable ω. All this certainly lies within the definition of the improper Riemann integral given in Chapter 1. We thus have

$$G(\omega) = \int_{-\infty}^{\infty} g(t)e^{-i\omega t}dt$$

with

$$g(t) = \frac{1}{2\pi}\int_{-\infty}^{\infty} G(\omega)e^{i\omega t}d\omega.$$

This coincides precisely with the definition of Fourier Transform given at the beginning of this chapter, right down to where the factor of 2π occurs! As has already been said, this positioning of the factor 2π is somewhat arbitrary, but it important to be consistent, and where it is here gives the most convenient progression to Laplace Transforms as indicated earlier in this section and in the next chapter.

In getting the Fourier Transform pair (as g and G are called) we have lifted the restriction that $g(t)$ be a periodic function. We have done this at a price however in that the improper Riemann integrals must be convergent. As we have already stated, unfortunately this is not the case for a wide class of functions including elementary functions such as sine, cosine, and even the constant function. However this serious problem is overcome through the development of generalised functions such as Dirac's δ function (see Chapter 2).

6.3 Basic Properties of the Fourier Transform

There are as many properties of the Fourier Transform as there are of the Laplace Transform. These involve shift theorems, transforming derivatives, etc.

Figure 6.2: The square wave function $f(t)$, and its Fourier Transform $F(\omega)$.

but they are not so widely used simply due to the restrictions on the class of functions that can be transformed. Most of the applications lie in the fields of Electrical and Electronic Engineering which are full of the jumpy and impulse like functions to which Fourier Transforms are particularly suited. Here is a simple and quite typical example.

Example 6.2 *Calculate the Fourier Transform of the "top hat" or rectangular pulse function defined as follows:-*

$$f(t) = \begin{cases} A & |t| \leq T \\ 0 & |t| > T \end{cases}$$

where A is a constant (amplitude of the pulse) and T is a second constant (width of the pulse).

Solution Evaluation of the integral is quite straightforward and the details are as follows

$$F(\omega) = \int_{-\infty}^{\infty} f(t)e^{-i\omega t} dt$$

$$= \int_{-T}^{T} Ae^{-i\omega t} dt$$

$$= \left[-\frac{A}{i\omega}e^{-i\omega t} \right]_{-T}^{T}$$

$$F(\omega) = \frac{2A}{\omega} \sin(\omega T).$$

Mathematically this is routine and rather uninteresting. However the graphs of $f(t)$ and $F(\omega)$ are displayed side by side in Figure 6.2, and it is worth a little discussion.

The relationship between $f(t)$ and $F(\omega)$ is that between a function of time ($f(t)$) and the frequencies that this function (called a signal by engineers) contains, $F(\omega)$. The subject of spectral analysis is a large one and sections of it are devoted to the relationship between a spectrum (often called a power spectrum) of a signal and the signal itself. This subject has been particularly fast growing

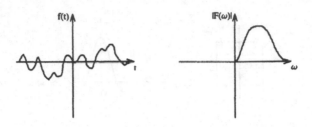

Figure 6.3: A typical wave form $f(t)$ and the amplitude of its Fourier Transform $|F(\omega)| = A(\omega)$.

since the days of the first satellite launch and the advent of satellite, then cable, and now digital television ensures its continuing growth. During much of the remainder of this chapter this kind of application will be hinted at, but a full account is of course not possible. The complex nature of $F(\omega)$ is not a problem. Most time series are not symmetric, so the modulus of $F(\omega)$ ($A(\omega)$) carries the frequency information.

A more typical-looking signal is shown on the left in Figure 6.3. Signals do not have a known functional form, and so their Fourier Transforms cannot be determined in closed form either. However some general characteristics are depicted on the right hand side of this figure. Only the modulus can be drawn in this form as the Fourier Transform is in general a complex quantity. The kind of shape $|F(\omega)|$ has is also fairly typical. High frequencies are absent as this would imply a rapidly oscillating signal; similarly very low frequencies are also absent as this would imply that the signal very rarely crossed the t axis. Thus the graph of $|F(\omega)|$ lies entirely between $\omega = 0$ and a finite value. Of course, any positive variation is theoretically possible between these limits, but the single maximum is most common. There is a little more to be said about these ideas in Section 6.5 when signal processing is discussed.

Sometimes it is inconvenient to deal with explicitly complex quantities, and the Fourier Transform is expressed in real and imaginary form as follows. If

$$F(\omega) = \int_{-\infty}^{\infty} f(t)e^{-i\omega t}dt$$

then

$$F_c(\omega) = \int_{0}^{\infty} f(t)\cos(\omega t)dt$$

is the Fourier Cosine Transform, and

$$F_s(\omega) = \int_{0}^{\infty} f(t)\sin(\omega t)dt$$

is the Fourier Sine Transform. We note that the bottom limits in both the Fourier Cosine and Sine Transform are zero rather than $-\infty$. This is in keeping with the notion that in practical applications t corresponds to time. Once more we warn that differences from these definitions involving positioning of the factor π are not uncommon. From the above definition it is easily deduced that

$$F(\omega) = \int_0^\infty [f(t) + f(-t)] \cos(\omega t) dt - i \int_0^\infty [f(t) - f(-t)] \sin(\omega t) dt$$

so if f is an odd function $[f(t) = -f(-t)]$, $F(\omega)$ is pure imaginary, and if f is an even function $[f(t) = f(-t)]$, $F(\omega)$ is real. We also note that if the bottom limit on each of the Fourier Sine and Cosine Transforms remained at $-\infty$ as in some texts, then the Fourier Sine Transform of an even function is zero as is the Fourier Cosine Transform of an odd function. This gives another good reason for the zero bottom limits for these transforms. Now let us examine some of the more common properties of Fourier Transforms, starting with the inverses of the Sine and Cosine Transforms. These are unsurprising: if

$$F_c(\omega) = \int_0^\infty f(t) \cos(\omega t) dt$$

then

$$f(t) = \frac{2}{\pi} \int_0^\infty F_c(\omega) \cos(\omega t) d\omega$$

and if

$$F_s(\omega) = \int_0^\infty f(t) \sin(\omega t) dt$$

then

$$f(t) = \frac{2}{\pi} \int_0^\infty F_s(\omega) \sin(\omega t) d\omega.$$

The proof of these is left as an exercise for the reader. The first property of these transforms we shall examine is their ability to evaluate certain improper real integrals in closed form. Most of these integrals are challenging to evaluate by other means (although readers familiar with the residue calculus also found in summary form in the next chapter should be able to do them). The following example illustrates this.

Example 6.3 *By considering the Fourier Cosine and Sine Transforms of the function* $f(t) = e^{-at}$, *a a constant, evaluate the two integrals*

$$\int_0^\infty \frac{\cos(kx)}{a^2 + x^2} dx \quad and \quad \int_0^\infty \frac{x \sin(kx)}{a^2 + x^2} dx.$$

Solution First of all note that the Cosine and Sine Transforms can be conveniently combined to give

$$\begin{aligned}
F_c(\omega) + iF_s(\omega) &= \int_0^\infty e^{(-a+i\omega)t} dt \\
&= \left[\frac{1}{-a+i\omega} e^{(-a+i\omega)t} \right]_0^\infty \\
&= \frac{1}{a - i\omega} = \frac{a + i\omega}{a^2 + \omega^2}
\end{aligned}$$

whence

$$F_c(\omega) = \frac{a}{a^2 + \omega^2} \text{ and } F_s(\omega) = \frac{\omega}{a^2 + \omega^2}.$$

Using the formula given for the inverse transforms gives

$$\frac{2}{\pi} \int_0^\infty \frac{a}{a^2 + \omega^2} \cos(\omega t) d\omega = e^{-at}$$

and

$$\frac{2}{\pi} \int_0^\infty \frac{\omega}{a^2 + \omega^2} \sin(\omega t) d\omega = e^{-at}.$$

Changing variables ω to x, t to k thus gives the results

$$\int_0^\infty \frac{\cos(kx)}{a^2 + x^2} dx = \frac{\pi}{2a} e^{-ak}$$

and

$$\int_0^\infty \frac{x \sin(kx)}{a^2 + x^2} dx = \frac{\pi}{2} e^{-ak}.$$

Laplace Transforms are extremely useful for finding the solution to differential equations. Fourier Transforms can also be so used; however the restrictions on the class of functions allowed is usually prohibitive. Assuming that the improper integrals exist, which requires that $f \to 0$ as $t \to \pm\infty$, let us start with the definition of the Fourier Transform

$$F(\omega) = \int_{-\infty}^\infty f(x) e^{-i\omega x} dx.$$

Since both limits are infinite, and the above conditions on f hold, we have that the Fourier Transform of $f'(t)$, the first derivative of $f(t)$, is straightforwardly $i\omega F(\omega)$ using integration by parts. In general

$$\int_{-\infty}^\infty \frac{d^n f}{dt^n} e^{-i\omega t} dt = (i\omega)^n F(\omega).$$

The other principal disadvantage of using Fourier Transform is the presence of i in the transform of odd derivatives and the difficulty in dealing with boundary conditions. Fourier Sine and Cosine Transforms are particularly useful however, for dealing with second order derivatives. The results

$$\int_0^\infty \frac{d^2 f}{dt^2} \cos(\omega t) dt = -\omega^2 F_c(\omega) - f'(0)$$

and

$$\int_0^\infty \frac{d^2 f}{dt^2} \sin(\omega t) dt = -\omega^2 F_s(\omega) + \omega f(0)$$

can be easily derived. These results are difficult to apply to solve differential equations of simple harmonic motion type because the restrictions imposed on

$f(t)$ are usually incompatible with the character of the solution. In fact, even when the solution *is* compatible, solving using Fourier Transforms is often impractical.

Fourier Transforms do however play an important role in solving partial differential equations as will be shown in the next section. Before doing this, we need to square up to some problems involving infinity. One of the common mathematical problems is coping with operations like integrating and differentiating when there are infinities around. When the infinity occurs in the limit of an integral, and there are questions as to whether the integral converges then we need to be particularly careful. This is certainly the case with Fourier Transforms of all types; it is less of a problem with Laplace Transforms. The following theorem is crucial in this point.

Theorem 6.4 (Lebesgue Dominated Convergence Theorem) *Let*

$$f_h, \quad h \in \mathcal{R}$$

be a family of piecewise continuous functions. If

1. *There exists a function g such that $|f_h(x)| \le g(x)$ $\forall x \in \mathcal{R}$ and all $h \in \mathcal{R}$.*

2.

$$\int_{-\infty}^{\infty} g(x)dx < \infty.$$

3.

$$\lim_{h \to 0} f_h(x) = f(x) \text{ for every } x \in \mathcal{R}$$

then

$$\lim_{h \to 0} \int_{-\infty}^{\infty} f_h(x)dx = \int_{-\infty}^{\infty} f(x)dx.$$

This theorem essentially tidies up where it is allowable to exchange the processes of taking limits and infinite integration. To see how important and perhaps counter intuitive this theorem is, consider the following simple example. Take a rectangular pulse or "top hat" function defined by

$$f_n(x) = \begin{cases} 1 & n \le x \le n+1 \\ 0 & \text{otherwise.} \end{cases}$$

Let $f(x) = 0$ so that it is true that $\lim_{n \to \infty} f_n(x) = f(x)$ for all $x \in \mathcal{R}$. However by direct integration it is obvious that

$$\lim_{h \to 0} \int_{-\infty}^{\infty} f_n(x)dx = 1 \neq 0 = \int_{-\infty}^{\infty} \lim_{n \to \infty} f_n(x)dx.$$

Thus the theorem does not hold for this function, and the reason is that the function $g(x)$ does not exist. In fact the improper nature of the integrals is incidental.

The proof of the Lebesgue Dominated Convergence Theorem involves classical analysis and is beyond the scope of this book. Suffice it to say that the function $g(x)$ which is integrable over $(-\infty, \infty)$ "dominates" the functions $f_h(x)$, and without it the limit and integral signs cannot be interchanged. As far as we are concerned, the following theorem is closer to home.

Denote by $G\{\mathcal{R}\}$ the family of functions defined on $\mathcal{R}$ with values in $\mathcal{C}$ which are piecewise continuous and absolutely integrable. These are essentially the Fourier Transforms as defined in the beginning of Section 6.2. For each $f \in G\{\mathcal{R}\}$ the Fourier Transform of f is defined for all $\omega \in \mathcal{R}$ by

$$F(\omega) = \int_{-\infty}^{\infty} f(x)e^{-i\omega x} dx$$

as in Section 6.2. That $G\{\mathcal{R}\}$ is a linear space over $\mathcal{C}$ is easy to verify. We now state the theorem.

Theorem 6.5 *For each $f \in G\{\mathcal{R}\}$,*

1. *$F(\omega)$ is defined for all $\omega \in \mathcal{R}$.*

2. *F is a continuous function on $\mathcal{R}$.*

3.

$$\lim_{\omega \pm \infty} F(\omega) = 0.$$

Proof To prove this theorem, we first need to use the previous Lebesgue Dominated Convergence Theorem. This was in fact the principal reason for stating it! We know that $|e^{i\omega x}| = 1$, hence

$$\int_{-\infty}^{\infty} |f(x)e^{-i\omega x}| dx = \int_{-\infty}^{\infty} |f(x)| dx < \infty.$$

Thus we have proved the first statement of the theorem; $F(\omega)$ is well defined for every real ω. This follows since the above equation tells us that $f(x)e^{-i\omega x}$ is absolutely integrable on $\mathcal{R}$ for each real ω. In addition, $f(x)e^{-i\omega x}$ is piecewise continuous and so belongs to $G\{\mathcal{R}\}$.

To prove that $F(\omega)$ is continuous is a little more technical. Consider the difference

$$F(\omega + h) - F(\omega) = \int_{-\infty}^{\infty} [f(x)e^{-i\omega(x+h)} - f(x)e^{-i\omega x}] dx$$

so

$$F(\omega + h) - F(\omega) = \int_{-\infty}^{\infty} f(x)e^{-i\omega x}[e^{-i\omega h} - 1] dx.$$

If we let

$$f_h(x) = f(x)e^{-i\omega x}[e^{-i\omega h} - 1]$$

to correspond to the $f_h(x)$ in the Lebesgue Dominated Convergence Theorem, we easily show that

$$\lim_{h \to 0} f_h(x) = 0 \text{ for every } x \in \mathcal{R}.$$

Also, we have that

$$|f_h(x)| = |f(x)||e^{-i\omega x}||e^{-i\omega h} - 1| \leq 2|f(x)|.$$

Now, the function $g(x) = 2|f(x)|$ satisfies the conditions of the function $g(x)$ in the Lebesgue Dominated Convergence Theorem, hence

$$\lim_{h \to 0} \int_{-\infty}^{\infty} f_h(x) dx = 0$$

whence

$$\lim_{h \to 0} [F(\omega + h) - F(\omega)] = 0$$

which is just the condition that the function $F(\omega)$ is continuous at every point of $\mathcal{R}$.

The last part of the theorem

$$\lim_{\omega \to \pm\infty} F(\omega) = 0$$

follows from the Riemann–Lebesgue lemma (see Theorem 4.2) since

$$F(\omega) = \int_{-\infty}^{\infty} f(x) e^{-i\omega x} dx = \int_{-\infty}^{\infty} f(x) \cos(\omega x) dx - i \int_{-\infty}^{\infty} f(x) \sin(\omega x) dx,$$
$$\lim_{\omega \to \pm\infty} F(\omega) = 0$$

is equivalent to

$$\lim_{\omega \to \pm\infty} \int_{-\infty}^{\infty} f(x) \cos(\omega x) dx = 0$$

and

$$\lim_{\omega \to \pm\infty} \int_{-\infty}^{\infty} f(x) \sin(\omega x) dx = 0$$

together. These two results are immediate from the Riemann–Lebesgue lemma (see Exercise 5 in Section 6.6). This completes the proof.

□

The next result, also stated in the form of a theorem, expresses a scaling property. There is no Laplace Transform equivalent due to the presence of zero in the lower limit of the integral in this case, but see Exercise 7 in Chapter 1.

Theorem 6.6 *Let $f(x) \in G\{\mathcal{R}\}$ and $a, b \in \mathcal{R}$, $a \neq 0$ and denote the Fourier Transform of f by $\mathcal{F}(f)$ so*

$$\mathcal{F}(f) = \int_{-\infty}^{\infty} f(x)e^{-i\omega x} dx.$$

Let $g(x) = f(ax + b)$. Then

$$\mathcal{F}(g) = \frac{1}{|a|} e^{i\omega b/a} \mathcal{F}(f) \left(\frac{\omega}{a} \right).$$

Proof As is usual with this kind of proof, the technique is simply to evaluate the Fourier Transform using its definition. Doing this, we obtain

$$\mathcal{F}(g(\omega)) = \int_{-\infty}^{\infty} f(ax + b)e^{-i\omega x} dx.$$

Now simply substitute $t = ax + b$ to obtain, if $a > 0$

$$\mathcal{F}(g(\omega)) = \frac{1}{a} \int_{-\infty}^{\infty} f(t)e^{-i\omega(t-b)/a} dt$$

and if $a < 0$

$$\mathcal{F}(g(\omega)) = -\frac{1}{a} \int_{-\infty}^{\infty} f(t)e^{-i\omega(t-b)/a} dt.$$

So, putting these results together we obtain

$$\mathcal{F}(g(\omega)) = \frac{1}{|a|} e^{i\omega b/a} \mathcal{F}(f) \left(\frac{\omega}{a} \right)$$

as required.

$\square$

The proofs of other properties follow along similar lines, but as been mentioned several times already the Fourier Transform applies to a restricted set of functions with a correspondingly smaller number of applications.

6.4 Fourier Transforms and Partial Differential Equations

In Chapter 5, two types of partial differential equation, parabolic and hyperbolic, were solved using Laplace Transforms. It was noted that Laplace Transforms were not suited to the solution of elliptic partial differential equations. Recall the reason for this. Laplace Transforms are ideal for solving initial value problems, but elliptic PDEs usually do not involve time and their solution does not yield evolutionary functions. Perhaps the simplest elliptic PDE is Laplace's equation ($\nabla^2 \phi = 0$) which, together with ϕ or its normal derivative given at the boundary gives a boundary value problem. The solutions are neither periodic nor are they initial value problems. Fourier Transforms as defined so far require that variables tend to zero at $\pm\infty$ and these are often natural assumptions for elliptic equations. There are also two new features that will now be introduced before problems are solved. First of all, partial differential equations such as $\nabla^2\phi = 0$ involve identical derivatives in x, y and, for three dimensional ∇^2, z too. It is logical therefore to treat them all in the same way. This leads to having to define two and three dimensional Fourier Transforms. Consider the function $\phi(x,y)$, then let

$$\hat{\phi}(k,y) = \int_{-\infty}^{\infty} \phi(x,y)e^{ikx}\,dx$$

and

$$\phi_F(k,l) = \int_{-\infty}^{\infty} \hat{\phi}(k,y)e^{ily}\,dy$$

so that

$$\phi_F(k,l) = \int_{-\infty}^{\infty}\int_{-\infty}^{\infty} \phi(x,y)e^{i(kx+ly)}\,dxdy$$

becomes a double Fourier Transform. In order for these transforms to exist, ϕ must tend to zero uniformly for (x,y) being a large distance from the origin, i.e. as $\sqrt{x^2 + y^2}$ becomes very large. The three dimensional Fourier Transform is defined analogously as follows.

$$\phi_F(k,l,m) = \int_{-\infty}^{\infty}\int_{-\infty}^{\infty}\int_{-\infty}^{\infty} \phi(x,y,z)e^{i(kx+ly+mz)}\,dxdydz.$$

With all these infinities around, the restrictions on ϕ_F are severe and applications are therefore limited. The frequency ω has been replaced by a three dimensional space (k,l,m) called phase space. However, the kind of problem that gives rise to Laplace's equation does fit this restriction, for example the behaviour of membranes, water or electromagnetic potential when subject to a point disturbance. For this kind of problem, the variable ϕ dies away to zero the further it is from the disturbance, therefore there is a good chance that the above infinite double or triple integral could exist. More useful for practical purposes however are problems in the finite domain and it is these that can be tackled usefully with a modification of the Fourier Transform. The unnatural part of the Fourier Transform is the imposition of conditions at infinity, and the modifications hinted at above have to do with replacing these by conditions at finite values. We therefore introduce the finite Fourier Transform (not to be confused with the FFT – Fast Fourier Transform). This is introduced in one dimension for clarity; the finite Fourier Transforms for two and three dimensions follow almost at once. If x is restricted to lie between say a and b, then the appropriate Fourier type transformation would be

$$\int_a^b \phi(x)e^{-ikx}\,dx.$$

This would then be applied to a problem in engineering or applied science where $a \leq x \leq b$. The two-dimensional version could be applied to a rectangle

$$a \leq x \leq b, \quad c \leq y \leq d$$

and is defined by

$$\int_c^d \int_a^b \phi(x,y)e^{-ikx-ily}dxdy.$$

Apart from the positive aspect of eliminating problems of convergence for (x,y) very far from the origin, the finite Fourier Transform unfortunately brings a host of negative aspects too. The first difficulty lies with the boundary conditions that have to be satisfied which unfortunately are now no longer infinitely far away. They can and do have considerable influence on the solution. One way to deal with this could be to cover the entire plane with rectangles with $\phi(x,y)$ being doubly periodic, i.e.

$$\phi(x,y) = \phi(x + N(b-a), y + M(d-c)), \quad M, N \text{ integers}$$

then revert to the original infinite range Fourier Transform. However, this brings problems of convergence and leads to having to deal with generalised functions. We shall not pursue this kind of finite Fourier Transform further here; but for a slightly different approach, see the next section. There seems to be a reverting to Fourier series here, after all the transform was obtained by a limiting process from series, and at first sight, all we seem to have done is reverse the process. A closer scrutiny reveals crucial differences, for example the presence of the factor $1/2\pi$ (or in general $1/l$ where l is the period) in front of the integrals for the Fourier series coefficients a_n and b_n. Much of the care and attention given to the process of letting the limit become infinite involved dealing with the zero that this factor produces. Finite Fourier Transforms do not have this factor. We return to some more of these differences in the next section, meanwhile let us do an example.

Example 6.7 *Find the solution to the two-dimensional Laplace Equation*

$$\frac{\partial^2 \phi}{\partial x^2} + \frac{\partial^2 \phi}{\partial y^2} = 0, \quad y > 0,$$

with

$$\frac{\partial \phi}{\partial x} \text{ and } \phi \to 0 \text{ as } \sqrt{x^2 + y^2} \to \infty, \quad \phi(x,0) = 1 \ |x| \leq 1 \ \phi(x,0) = 0, \ |x| > 1.$$

Use Fourier Transforms in x.

Solution Let

$$\bar{\phi}(k,y) = \int_{-\infty}^{\infty} \phi(x,y)e^{-ikx}dx$$

then

$$\int_{-\infty}^{\infty} \frac{\partial^2}{\partial y^2} e^{-ikx} dx = \frac{\partial^2}{\partial y^2} \left(\int_{-\infty}^{\infty} \phi e^{-ikx} dx \right)$$

$$= \frac{\partial^2 \bar{\phi}}{\partial y^2}.$$

Also

$$\int_{-\infty}^{\infty} \frac{\partial^2 \phi}{\partial x^2} e^{-ikx} dx$$

can be integrated by parts as follows

$$\int_{-\infty}^{\infty} \frac{\partial^2 \phi}{\partial x^2} e^{-ikx} dx = \left[\frac{\partial \phi}{\partial x} e^{-ikx} \right]_{-\infty}^{\infty} + ik \int_{-\infty}^{\infty} \frac{\partial \phi}{\partial x} e^{-ikx} dx$$

$$= ik \left[\phi e^{-ikx} \right]_{-\infty}^{\infty} - k^2 \int_{-\infty}^{\infty} \phi e^{-ikx} dx$$

$$= -k^2 \bar{\phi}.$$

We have used the conditions that both ϕ and its x derivative decay to zero as $x \to \pm\infty$. Hence if we take the Fourier Transform of the Laplace Equation in the question,

$$\int_{-\infty}^{\infty} \left(\frac{\partial^2 \phi}{\partial x^2} + \frac{\partial^2 \phi}{\partial y^2} \right) e^{-ikx} dx = 0$$

or

$$\frac{\partial^2 \bar{\phi}}{\partial y^2} - k^2 \bar{\phi} = 0.$$

As $\bar{\phi} \to 0$ for large y, the (allowable) solution is

$$\bar{\phi} = C e^{-|k|y}.$$

Now, we can apply the condition on $y = 0$

$$\bar{\phi}(k,0) = \int_{-\infty}^{\infty} \phi(x,0) e^{-ikx} dx$$

$$= \int_{-1}^{1} e^{-ikx} dx$$

$$= \frac{e^{ik} - e^{-ik}}{ik}$$

$$= \frac{2\sin(k)}{k}$$

whence

$$C = \frac{2\sin(k)}{k}$$

and

$$\bar{\phi} = \frac{2\sin(k)}{k}e^{-|k|y}.$$

In order to invert this, we need the Fourier Transform equivalent of the Convolution theorem (see Chapter 3). To see how this works for Fourier Transforms, consider the convolution of the two general functions f and g

$$
\begin{aligned}
F * G = \int_{-\infty}^{\infty} f(\tau)g(t-\tau)d\tau &= \int_{-\infty}^{\infty}\int_{-\infty}^{\infty} f(\tau)G(x)e^{-ix(t-\tau)}d\tau dx \\
&= \int_{-\infty}^{\infty} G(x)e^{-ixt}\int_{-\infty}^{\infty} f(\tau)e^{ix\tau}d\tau dx \\
&= \frac{1}{2\pi}\int_{-\infty}^{\infty} G(x)F(x)e^{-ixt}dx \\
&= \frac{1}{2\pi}\mathcal{F}(FG).
\end{aligned}
$$

Now, the Fourier Transform of $e^{-|k|y}$ is

$$\frac{y}{k^2+\omega^2}$$

and that of $\phi(x,0)$ is

$$\frac{2\sin(k)}{k},$$

hence by the convolution theorem,

$$
\begin{aligned}
\phi(x,y) &= \frac{1}{\pi}y\int_{-1}^{1}\frac{d\tau}{(x-\tau)^2+y^2} \\
&= \frac{1}{\pi}y\left(\tan^{-1}\left(\frac{x-1}{y}\right) + \tan^{-1}\left(\frac{x+1}{y}\right)\right)
\end{aligned}
$$

is the required solution.

6.5   Signal Processing

There is no doubt that the most prolific application of Fourier Transforms lies in the field of signal processing. As this is a branch of electrical engineering an in depth analysis is out of place here, but some discussion is helpful. To start with, we return to the complex form of the Fourier series and revisit explicitly the close connections between Fourier series and finite Fourier Transforms. From Chapter 4 (and Section 6.2)

$$f(x) = \sum_{n=-\infty}^{\infty} c_n e^{inx}$$

with

$$c_n = \frac{1}{2\pi} \int_{-\pi}^{\pi} f(x)e^{-inx}dx.$$

Thus $2\pi c_n$ is the finite Fourier Transform of $f(x)$ over the interval $[-\pi, \pi]$ and the "inverse" is the Fourier series for $f(x)$. If c_n is given as a sequence, then $f(x)$ is easily found. (In practice, the sequence c_n consists of but a few terms.) The resulting $f(x)$ is of course periodic as it is the sum of terms of the type $c_n e^{inx}$ for various n. Hence the finite Fourier Transform of this type of periodic function is a sequence of numbers, usually four or five at most. It is only a short step from this theoretical treatment of finite Fourier Transforms to the analysis of periodic signals of the type viewed on cathode ray oscilloscopes. A simple illustrative example follows.

Example 6.8 *Consider the simple "top hat" function $f(x)$ defined by*

$$f(x) = \begin{cases} 1 & x \in [0, \pi] \\ 0 & otherwise. \end{cases}$$

Find its finite Fourier Transform and finite Fourier Sine Transform.

Solution The finite Fourier Transform of this function is simply

$$\int_0^{\pi} e^{-inx}dx = \left[-\frac{e^{-inx}}{in}\right]_0^{\pi}$$

$$= \frac{1}{in}(1 - e^{-in\pi})$$

which can be written in a number of ways:-

$$\frac{1}{in}(1 - (-1)^n); \quad -\frac{2i}{(2k+1)}; \quad \frac{2\sin(\frac{n\pi}{2})}{n}e^{in\pi/2}.$$

The finite Sine Transform is a more natural object to find: it is

$$\int_0^{\pi} \sin(nx)dx = \frac{1}{n}(1 - (-1)^n) = \frac{2}{2k+1}, \quad n, k \text{ integers.}$$

Let us use this example to illustrate the transition from finite Fourier Transforms to Fourier Transforms proper. The inverse finite Fourier Transform of the function $f(x)$ as defined in Example 6.8 is the Fourier series

$$f(x) = \frac{1}{2\pi} \sum_{-\infty}^{\infty} \frac{1 - (-1)^n}{in}e^{inx} \quad 0 \le x \le \pi.$$

However, although $f(x)$ is only defined in the interval $[0, \pi]$, the Fourier series is periodic, period 2π. It therefore represents a square wave shown in Figure 6.4. Of course, $x \in [0, \pi]$ so $f(x)$ is represented as a "window" to borrow a phrase from time series and signal analysis. If we write $x = \pi t/l$, then let $l \to \infty$: we

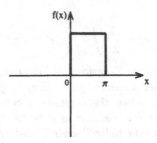

Figure 6.4: The square wave

regain the transformation that took us from Fourier series to Fourier Transforms proper, Section 6.2. However, what we have in Figure 6.5 is a typical signal. The Fourier Transform of this signal taken as a whole of course does not exist as conditions at $\pm\infty$ are not satisfied. In the case of an actual signal therefore, the use of the Fourier Transform is made possible by restricting attention to a window, that is a finite range of t. This gives rise to a series (Fourier series) representation of the Fourier Transform of the signal. This series has a period which is dictated by the (usually artificially generated) width of the window. The Fourier coefficients give important information on the frequencies present in the original signal. This is the fundamental reason for using these methods to examine signals from such diverse applications as medicine, engineering and seismology. Mathematically, the way forward is through the introduction of the Dirac-δ function as follows. We have that

$$\mathcal{F}\{\delta(t-t_0)\} = \int_{-\infty}^{\infty} \delta(t-t_0)e^{-i\omega t}dt = e^{-i\omega t_0}$$

and the inverse result implies that

$$\mathcal{F}\{e^{-it_0 t}\} = 2\pi\delta(\omega - \omega_0).$$

Whence we can find the Fourier Transform of a given Fourier series (written in complex exponential form) by term by term evaluation provided such operations are legal in terms of defining the Dirac-δ as the limiting case of an infinitely tall but infinitesimally thin rectangular pulse of unit area (se Section 2.6).

$$f(t) \sim \sum_{n=-\infty}^{\infty} F_n e^{in\omega_0 t}$$

so that

$$\mathcal{F}\{f(t)\} \sim \mathcal{F}\left\{\sum_{n=-\infty}^{\infty} F_n e^{in\omega_0 t}\right\}$$

$$= \sum_{n=-\infty}^{\infty} F_n \mathcal{F}\{e^{in\omega_0 t}\}$$

which implies

$$\mathcal{F}\{f(t)\} \sim 2\pi \sum_{n=-\infty}^{\infty} F_n \delta(\omega - n\omega_0).$$

Now, suppose we let

$$f(t) = \sum_{n=-\infty}^{\infty} \delta(t - nT)$$

that is $f(t)$ is an infinite train of equally spaced Dirac-δ functions (called a Shah function by electrical and electronic engineers), then $f(t)$ is certainly periodic (of period T). Of course it is not piecewise continuous, but if we follow the limiting processes through carefully, we can find a Fourier series representation of $f(t)$ as

$$f(t) \sim \sum_{n=-\infty}^{\infty} F_n e^{-in\omega_0 t}$$

where $\omega_0 = 2\pi/T$, with

$$F_n = \frac{1}{T} \int_{-T/2}^{T/2} f(t) e^{-in\omega_0 t} dt$$

$$= \frac{1}{T} \int_{-T/2}^{T/2} \delta(t) e^{-in\omega_0 t} dt$$

$$= \frac{1}{T}$$

for all n. Hence we have the result that

$$\mathcal{F}\{f(t)\} = 2\pi \sum_{n=-\infty}^{\infty} \frac{1}{T} \delta(\omega - n\omega_0)$$

$$= \omega_0 \sum_{n=-\infty}^{\infty} \delta(\omega - n\omega_0).$$

Which means that the Fourier Transform of an infinite string of equally spaced Dirac-δ functions (Shah function) is another string of equally spaced Dirac-δ functions:

$$\mathcal{F}\left\{ \sum_{n=-\infty}^{\infty} \delta(t - nT) \right\} = \omega_0 \sum_{n=-\infty}^{\infty} \delta(\omega - n\omega_0).$$

It is this result that is used extensively by engineers and statisticians when analysing signals using sampling. Mathematically, it is of interest to note that with $T = 2\pi$ ($\omega_0 = 1$) we have found an invariant under Fourier Transformation.

If $f(t)$ and $F(\omega)$ are a Fourier Transform pair, then the quantity

$$E = \int_{-\infty}^{\infty} |f(t)|^2 dt$$

is called the *total energy*. This expression is an obvious carry over from $f(t)$ representing a time series. (Attempts at dimensional analysis are fruitless due to the presence of a dimensional one on the right hand side. This is annoying to physicists and engineers, very annoying!) The quantity $|F(\omega)|^2$ is called the *energy spectral density* and the graph of this against ω is called the *energy spectrum* and remains a very useful guide as to how the signal $f(t)$ can be thought of in terms of its decomposition into frequencies. The energy spectrum of $\sin(kt)$ for example is a single spike in ω space corresponding to the frequency $2\pi/k$. The constant energy spectrum where all frequencies are present in equal measure corresponds to the "white noise" signal characterised by a hiss when rendered audible. The two quantities $|f(t)|^2$ and energy spectral density are connected by the transform version of Parseval's theorem (sometimes called Rayleigh's theorem, or Plancherel's identity). See Theorem 4.19 for the series version.

Theorem 6.9 (Parseval's, for Transforms) *If $f(t)$ has a Fourier Transform $F(\omega)$ and*

$$\int_{-\infty}^{\infty} |f(t)|^2 dt < \infty$$

then

$$\int_{-\infty}^{\infty} |f(t)|^2 dt = \frac{1}{2\pi} \int_{-\infty}^{\infty} |F(\omega)|^2 d\omega.$$

Proof The proof is straightforward:-

$$\int_{-\infty}^{\infty} f(t)f^*(t)dt = \int_{-\infty}^{\infty} f(t)\frac{1}{2\pi}\int_{-\infty}^{\infty} F(\omega)e^{i\omega t}d\omega dt$$

$$= \frac{1}{2\pi}\int_{-\infty}^{\infty} F(\omega)\int_{-\infty}^{\infty} f(t)e^{i\omega t}dt d\omega$$

$$= \frac{1}{2\pi}\int_{-\infty}^{\infty} F(\omega)(F(\omega))^* d\omega$$

$$= \frac{1}{2\pi}\int_{-\infty}^{\infty} |F(\omega)|^2 d\omega$$

where f^* is the complex conjugate of $f(t)$. The exchange of integral signs is justified as long as their values remain finite.

□

Most of the applications of this theorem lie squarely in the field of signal processing, but here is a simple example using the definition of energy spectral density.

Example 6.10 *Determine the energy spectral densities for the following functions:*

(i)

$$f(t) = \begin{cases} A & |t| < T \\ 0 & otherwise \end{cases}$$

This is the same function as in Example 6.2.

(ii)

$$f(t) = \begin{cases} e^{-at} & t \ge 0 \\ 0 & t < 0. \end{cases}$$

Solution The energy spectral density $|F(\omega)|^2$ is found from $f(t)$ by first finding its Fourier Transform. Both calculations are essentially routine.

(i)

$$\begin{aligned} F(\omega) &= \int_{-\infty}^{\infty} f(t) e^{i\omega t} dt \\ &= A \int_{-T}^{T} e^{i\omega t} dt \\ &= \frac{A}{i\omega} \left[e^{i\omega t} \right]_{-T}^{T} \\ &= \frac{A}{i\omega} [e^{i\omega T} - e^{-i\omega T}] = \frac{2A \sin(\omega t)}{\omega} \end{aligned}$$

as already found in Example 6.2. So we have that

$$|F(\omega)|^2 = \frac{4A^2 \sin^2(\omega T)}{\omega^2}.$$

(ii)

$$\begin{aligned} F(\omega) &= \int_0^{\infty} e^{-at} e^{-i\omega t} dt \\ &= \left[-\frac{e^{(-a-i\omega)t}}{a+i\omega} \right]_0^{\infty} \\ &= \frac{a-i\omega}{a^2+\omega^2}. \end{aligned}$$

Hence

$$|F(\omega)|^2 = \frac{a-i\omega}{a^2+\omega^2} \cdot \frac{a+i\omega}{a^2+\omega^2} = \frac{1}{a^2+\omega^2}.$$

There is another aspect of signal processing that ought to be mentioned. Most signals are not deterministic and have to be analysed by using statistical techniques such as sampling. It is by sampling that a time series which is given in the form of an analogue signal (a wiggly line as in Figure 6.5) is transformed

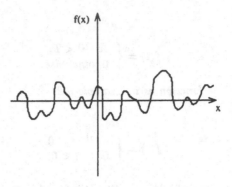

Figure 6.5: A time series

into a digital one (usually a series of zeros and ones). The Fourier Transform is a means by which a signal can be broken down into component frequencies, from t space to ω space. This cannot be done directly since time series do not conveniently obey the conditions at $\pm\infty$ that enable the Fourier Transform to exist formally. The autocovariance function is the convolution of f with itself and is a measure of the agreement (or correlation) between two parts of the signal time t apart. It turns out that the autocovariance function of the time series is, however well behaved at infinity and it is usually this function that is subject to spectral decomposition either directly (analogue) or via sampling (digital). Digital time series analysis is now a very important subject due to the omnipresent (digital) computer. We will all experience digital television signals in the next few years; this is the latest manifestation of digital signal analysis. We shall not pursue it further here. What we hope to have achieved here is an appreciation of the importance of Fourier Transforms and Fourier series to the subject.

Let us finish this chapter by doing an example which demonstrates a slightly different way of illustrating the relationship between finite and standard Fourier Transforms.

The electrical engineering fraternity define the window function by $W(x)$ where

$$W(x) = \begin{cases} 0 & |x| > \frac{1}{2} \\ \frac{1}{2} & |x| = \frac{1}{2} \\ 1 & |x| < \frac{1}{2}. \end{cases}$$

giving the picture of Figure 6.6. This is almost the same as the "top hat" function defined in the last example. Spot the subtle difference.

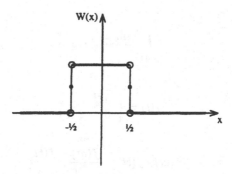

Figure 6.6: The window function $W(x)$

Use of this function in Fourier Transforms immediately converts a Fourier Transform into a finite Fourier Transform as follows

$$\int_{-\infty}^{\infty} W\left(\frac{x - \frac{1}{2}(b+a)}{b-a}\right) f(x)e^{-i\omega x}dx = \int_{a}^{b} f(x)e^{-i\omega x}dx.$$

It is easy to check that if

$$t = \frac{x - \frac{1}{2}(b+a)}{b-a}$$

then $t > \frac{1}{2}$ corresponds to $x > b$ and $t < -\frac{1}{2}$ corresponds to $x < a$. What this approach does is to move the work from inverting a finite Fourier Transform in terms of Fourier series to evaluating

$$\frac{1}{2\pi}\int_{-\infty}^{\infty} F_W(\omega)e^{i\omega x}d\omega$$

where $F_W(\omega)$ is the Fourier Transform of the "windowed" version of $f(x)$. It will come as no surprise to learn that the calculation of this integral is every bit as difficult (or easy) as directly inverting the finite Fourier Transform. The choice lies between working with Fourier series directly or working with $F_W(\omega)$ which involves series of generalised functions.

6.6 Exercises

1. Determine the Fourier Transform of the function $f(t)$ defined by

$$f(t) = \begin{cases} k & -T \le t < 0 \\ -k & 0 \le t < T \\ 0 & \text{otherwise.} \end{cases}$$

2. If $f(t) = e^{-t^2}$, find its Fourier Transform.

3. Show that

$$\int_0^\infty \frac{\sin(u)}{u} du = \frac{\pi}{2}.$$

4. Define

$$f(t) = \begin{cases} e^{-t} & t \geq 0 \\ 0 & t < 0 \end{cases}$$

and show that

$$f(at) * f(bt) = \frac{f(at) - f(bt)}{b - a}$$

where a and b are arbitrary real numbers. Hence also show that $f(at) *$ $f(at) = tf(at)$ where $*$ is the Fourier Transform version of the convolution operation defined by

$$f(t) * g(t) = \int_{-\infty}^{\infty} f(\tau)g(t - \tau)d\tau.$$

5. Consider the integral

$$g(x) = \int_{-\infty}^{\infty} f(t)e^{-2\pi i x t} dt$$

and, using the substitution $u = t - 1/2x$, show that $|g(x)| \to 0$ hence providing a simple illustration of the Riemann–Lebesgue lemma.

6. Derive Parseval's formula:-

$$\int_{-\infty}^{\infty} f(t)G(it)dt = \int_{-\infty}^{\infty} F(it)g(t)dt$$

where F and G are the Fourier Transforms of $f(t)$ and $g(t)$ respectively and all functions are assumed to be well enough behaved.

7. Define $f(t) = 1 - t^2$, $-1 < t < 1$, zero otherwise, and $g(t) = e^{-t}$, $0 \leq t < \infty$, zero otherwise. Find the Fourier Transforms of each of these functions and hence deduce the value of the integral

$$\int_0^\infty \frac{4e^{-t}}{t^3}(t\cosh(t) - \sinh(t))dt$$

by using Parseval's formula (see Exercise 6). Further, use Parseval's theorem for Fourier Transforms, Theorem 6.9, to evaluate the integral

$$\int_0^\infty \frac{(t\cos(t) - \sin(t))^2}{t^6}dt$$

8. Consider the partial differential equation

$$u_t = ku_{xx} \quad x > 0, \ t > 0$$

with boundary conditions $u(x,0) = g(x)$, $u(0,t) = 0$, where $g(x)$ is a suitably well behaved function of x. Take Laplace Transforms in t to obtain an ordinary differential equation, then take Fourier Transforms in x to solve this in the form of an improper integral.

9. The Helmholtz equation takes the form $u_{xx} + u_{yy} + k^2 u = f(x,y)$ $-\infty < x, y < \infty$. Assuming that the functions $u(x,y)$ and $f(x,y)$ have Fourier Transforms show that the solution to this equation can formally be written:-

$$u(x,y) = -\frac{1}{4\pi^2} \int \int \int \int e^{-i[\lambda(x-\xi)+\mu(y-\eta)]} \frac{f(\xi,\eta)}{\lambda^2 + \mu^2 - k^2} d\lambda d\mu d\xi d\eta,$$

where all the integrals are from $-\infty$ to ∞. State carefully the conditions that must be obeyed by the functions $u(x,y)$ and $f(\xi,\eta)$.

10. Use the window function $W(x)$ defined in the last section to express the coefficients (c_n) in the complex form of the Fourier series for an arbitrary function $f(x)$ in terms of an integral between $-\infty$ and ∞. Hence, using the inversion formula for the Fourier Transform find an integral for $f(x)$ Compare this with the Fourier series and the derivation of the transform in Section 6.2, noting the role of periodicity and the window function.

11. The two series

$$F_k = \sum_{n=0}^{N-1} f_n e^{-ink\Delta\omega T},$$

$$f_n = \frac{1}{N} \sum_{k=0}^{N-1} F_k e^{ink\Delta\omega T}$$

where $\Delta\omega = 2\pi/NT$ define the *discrete* Fourier Transform and its inverse. Outline how this is derived from continuous (standard) Fourier Transforms by considering the series f_n as a sampled version of the time series $f(t)$.

12. Using the definition in the previous exercise, determine the discrete Fourier Transform of the sequence $\{1, 2, 1\}$ with $T = 1$.

13. Find, in terms of Dirac-δ functions the Fourier Transforms of $\cos(\omega_0 t)$ and $\sin(\omega_0 t)$, where ω_0 is a constant.

$$7$$

Complex Variables and Laplace Transforms

7.1 Introduction

The material in this chapter is written on the assumption that you have some
familiarity with complex variable theory (or complex analysis). That is we
assume that defining $f(z)$ where $z = x + iy$, $i = \sqrt{-1}$, and where x and y
are independent variables is not totally mysterious. In Laplace Transforms, s
can fruitfully be thought of as a complex variable. Indeed parts of this book
(Section 6.2 for example) have already strayed into this territory.

For those for whom complex analysis is entirely new (rather unlikely if you
have got this far!), there are many excellent books on the subject. Those by
Priestley (1985), Stewart and Tall (1983) or (more to my personal taste) Needham (1997) or Osborne (1999) are recommended. We give a brief resumé of
required results without detailed proof. The principal reason for needing complex variable theory is to be able to use and understand the proof of the formula
for inverting the Laplace Transform. Section 7.6 goes a little further than this,
but not much. The complex analysis given here is therefore by no means complete.

7.2 Rudiments of Complex Analysis

In this section we shall use $z(= x + iy)$ as our complex variable. It will be
assumed that complex numbers, e.g. $2 + 3i$ are familiar, as are their representation on an Argand diagram (Figure 7.1). The quantity $z = x + iy$ is different
in that x and y are both (real) variables, so z must be a complex variable. In

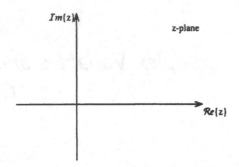

Figure 7.1: The Argand diagram

this chapter we shall be concerned with $f(z)$, that is the properties of functions of a complex variable.

In general we are able to split $f(z)$ into real and imaginary parts

$$f(z) = \phi(x, y) + i\psi(x, y)$$

where ϕ and ψ are real functions of the two real variables x and y. For example

$$z^2 = x^2 - y^2 + i(2xy)$$

$$\sin(z) = \sin(x)\cosh(y) + i\sinh(x)\cos(y).$$

This is the reason that many results from two real variable calculus are useful in complex analysis. However it does not by any means tell the whole story of the power of complex analysis.

One astounding fact is that if $f(z)$ is a function that is once differentiable with respect to z, then it is also differentiable twice, three times, as many times as we like! The function $f(z)$ is then called a *regular* or *analytic* function. Such a concept is absent in real analysis where differentiability to any specific order does not guarantee further differentiability. If $f(z)$ is regular then we can show that

$$\frac{\partial \phi}{\partial x} = -\frac{\partial \psi}{\partial y} \text{ and } \frac{\partial \phi}{\partial y} = \frac{\partial \psi}{\partial x}.$$

These are called the Cauchy–Riemann equations. Further, we can show that

$$\nabla^2 \phi = 0, \text{ and } \nabla^2 \psi = 0$$

i.e. both ϕ and ψ are harmonic. We shall not prove any of these results.

Once a function of a single real variable is deemed many times differentiable in a certain range (one dimensional domain), then one possibility is to express the function as a power series. Power series play a central role in both real and complex analysis. The power series of a function of a real variable about the point $x = x_0$ is the Taylor series. Truncate this series after $n + 1$ terms and we get the result

$$f(x) = f(x_0) + (x - x_0)f'(x_0) + \frac{(x - x_0)^2}{2!}f''(x_0) + \cdots + \frac{(x - x_0)^n}{n!}f^{(n)}(x_0) + R_n$$

where

$$R_n = \frac{(x - x_0)^{n+1}}{(n + 1)!}f^{(n+1)}(x_0 + \theta(x - x_0)) \quad (0 \le \theta \le 1)$$

is the remainder. This series is valid in the range $|x - x_0| \le r$, r is called the radius of convergence. The function $f(x)$ is differentiable $n + 1$ times in the region $x_0 - r \le x \le x_0 + r$.

This result carries through unchanged to complex variables

$$f(z) = f(z_0) + (z - z_0)f'(z_0) + \frac{(z - z_0)^2}{2!}f''(z_0) + \cdots + \frac{(z - z_0)^n}{n!}f^{(n)}(z_0) + \cdots$$

but now of course $f(z)$ is analytic in a *disc* $|z - z_0| < r$. What follows, however has no direct analogy in real variables. The next step is to consider functions that are regular in an *annular* region $r_1 < |z - z_0| < r_2$, or in the limit, a *punctured disc* $0 < |z - z_0| < r_2$. In such a region, a complex function $f(z)$ possesses a Laurent series

$$f(z) = \sum_{n=-\infty}^{\infty} a_n(z - z_0)^n.$$

The $a_0 + a_1(z - z_0) + a_2(z - z_0)^2 + \cdots$ part of the series is the "Taylor part" but the a_ns cannot in general be expressed in terms of derivatives of $f(z)$ evaluated at $z = z_0$ for the very good reason that $f(z)$ is not analytic at $z = z_0$ so such derivatives are not defined! The rest of the Laurent series

$$\cdots + \frac{a_{-2}}{(z - z_0)^2} + \frac{a_{-1}}{(z - z_0)}$$

is called the principal part of the series and is usefully employed in helping to characterise with some precision what happens to $f(z)$ at $z = z_0$. If $a_{-1}, a_{-2}, \ldots, a_{-n}, \ldots$ are all zero, then $f(z)$ *is* analytic at $z = z_0$ and possesses a Taylor series. For such functions which have little application here, $a_1 = f'(z_0), a_2 = \frac{1}{2!}f''(z_0)$ etc. The last coefficient of the principal part, a_{-1} the coefficient of $\frac{1}{z - z_0}$, is termed the *residue* of $f(z)$ at $z = z_0$ and has particular importance. Dealing with finding the a_ns (positive or negative n) in the Laurent series for $f(z)$ takes us into the realm of complex integration. This is because, in order to find a_n we manipulate the values $f(z)$ has in the region of validity of the Laurent series, i.e. in the annular region, to infer something about a_n. To do this, a complex version of the mean value theorem for integrals is used called Cauchy's integral formulae. In general, these take the form

$$g^{(n)}(z_0) = \frac{n!}{2\pi i}\int_C \frac{g(z)}{(z - z_0)^{n+1}}dz$$

where we assume that $g(z)$ is an analytic function of z everywhere inside C (even at $z = z_0$). (Integration in this form is formally defined in the next section.) The case $n = 1$ gives insight, as the integral then assumes classic mean value form. Now if we consider $f(z)$, which is not analytic at $z = z_0$, and let

$$f(z) = \frac{g(z)}{(z - z_0)^n}$$

where $g(z)$ *is* analytic at $z = z_0$, then we have said something about the behaviour of $f(z)$ at $z = z_0$. $f(z)$ has what is called a *singularity* at $z = z_0$, and we have specified this as a pole of order n. A consequence of $g(z)$ being analytic and hence possessing a Taylor series about $z = z_0$ valid in $|z - z_0| = r$ is that the principal part of $f(z)$ has leading term $\dfrac{a_{-n}}{(z - z_0)^n}$ with a_{-n-1}, a_{-n-2} etc. all equalling zero. A pole of order n is therefore characterised by $f(z)$ having n terms in its principal part. If $n = 1$, $f(z)$ has a simple pole (pole of order 1) at $z = z_0$. If there are infinitely many terms in the principal part of $f(z)$, then $f(z)$ is called an *essential* singularity of $f(z)$, and there are no Cauchy integral formulae. The foregoing theory is not valid for such functions. It is also worth mentioning branch points at this stage, although they do not feature for a while yet. A branch point of a function is a point at which the function is many valued. $\sqrt{z}$ is a simple example, $\text{Ln}(z)$ a more complicated one. The ensuing theory is not valid for branch points.

7.3   Complex Integration

The integration of complex functions produces many surprising results, none of which are even hinted at by the integration of a single real variable. Integration in the complex z plane is a line integral. Most of the integrals we shall be concerned with are integrals around closed contours.

Suppose $p(t)$ is a complex valued function of the real variable t, with $t \in [a, b]$ which has real part $p_r(t)$ and imaginary part $p_i(t)$ both of which are piecewise continuous on $[a, b]$. We can then integrate $p(t)$ by writing

$$\int_a^b p(t)dt = \int_a^b p_r(t)dt + i \int_a^b p_i(t)dt.$$

The integrals on the right are Riemann integrals. We can now move on to define contour integration. It is possible to do this in terms of line integrals. However, line integrals have not made an appearance in this book so we avoid them here. Instead we introduce the idea of contours via the following set of simple definitions:

Definition 7.1   *1. An arc C is a set of points $\{(x(t), y(t)) : t \in [a, b]\}$ where $x(t)$ and $y(t)$ are continuous functions of the real variable t. The arc is conveniently described in terms of the complex valued function z of the real variable t where*

$$z(t) = x(t) + iy(t).$$

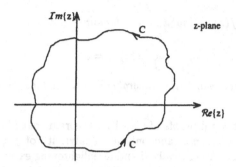

Figure 7.2: The curve C in the complex plane

2. *An arc is* simple *if it does not cross itself. That is* $z(t_1) = z(t_2) \Rightarrow t_1 = t_2$ *for all* $t_1, t_2 \in [a, b]$.

3. *An arc is* smooth *if* $z'(t)$ *exists and is non-zero for* $t \in [a, b]$. *This ensures that C has a continuously turning tangent.*

4. *A (simple)* contour *is an arc consisting of a finite number of (simple) smooth arcs joined end to end. When only the initial and final values of* $z(t)$ *coincide, the contour is a (simple) closed contour.*

Here, all contours will be assumed to be simple. It can be seen that as t varies, $z(t)$ describes a curve (contour) on the z-plane. t is a real parameter, the value of which uniquely defines a point on the curve $z(t)$ on the z-plane. The values $t = a$ and $t = b$ give the end points. If they are the same as in Figure 7.2 the contour is closed, and it is closed contours that are of interest here. By convention, t increasing means that the contour is described anti-clockwise. This comes from the parameter θ describing the circle $z(\theta) = e^{i\theta} = \cos(\theta) + i\sin(\theta)$ anti-clockwise as θ increases. Here is the definition of integration which follows directly from this parametric description of a curve.

Definition 7.2 *Let C be a simple contour as defined above extending from the point* $\alpha = z(a)$ *to* $\beta = z(b)$. *Let a domain* $\mathcal{D}$ *be a subset of the complex plane and let the curve C lie wholly within it. Define* $f(z)$ *to be a piecewise continuous function on C, that is* $f(z(t))$ *is piecewise continuous on the interval* $[a, b]$. *Then the contour integral of* $f(z)$ *along the contour C is*

$$\int_C f(z)dz = \int_a^b f(z)dz = \int_a^b f(z(t))z'(t)dt.$$

In addition to $f(z)$ being continuous at all points of a curve C (see Figure 7.2) it will be also be assumed to be of finite length (rectifiable).

An alternative definition following classic Riemann integration (see Chapter 1) is also possible.

The first result we need is Cauchy's theorem which we now state.

Theorem 7.3 *If $f(z)$ is analytic in a domain D and on its (closed) boundary C then*

$$\oint_C f(z)dz = 0$$

where the small circle within the integral sign denotes that the integral is around a closed loop.

Proof There are several proofs of Cauchy's theorem that place no reliance on the analyticity of $f(z)$ *on* C and only use analyticity of $f(z)$ *inside* C. However these proofs are rather involved and unenlightening except for those whose principal interest is pure mathematics. The most straightforward proof makes use of Green's theorem in the plane which states that if $P(x,y)$ and $Q(x,y)$ are continuous and have continuous derivatives in D and on C, then

$$\oint_C Pdx + Qdy = \int\int_D \left(\frac{\partial Q}{\partial x} - \frac{\partial P}{\partial y}\right) dxdy.$$

This is easily proved by direct integration of the right hand side. To apply this to complex variables, consider $f(z) = P(x,y) + iQ(x,y)$ and $z = x + iy$ so

$$
\begin{aligned}
f(z)dz &= (P(x,y) + iQ(x,y))(dx + idy) \\
&= P(x,y)dx - Q(x,y)dy + i(P(x,y)dy + Q(x,y)dx)
\end{aligned}
$$

and so

$$\oint_C f(z)dz = \oint_C (P(x,y)dx - Q(x,y)dy) + i\oint_C (P(x,y)dy + Q(x,y)dx).$$

Using Green's theorem in the plane for both integrals gives

$$\oint_C f(z)dz = \int\int_D \left(-\frac{\partial Q}{\partial x} - \frac{\partial P}{\partial y}\right) dxdy + i\int\int_D \left(\frac{\partial P}{\partial x} - \frac{\partial Q}{\partial y}\right) dxdy.$$

Now the Cauchy–Riemann equations imply that if

$$f(z) = P(x,y) + iQ(x,y) \text{ then } \frac{\partial P}{\partial x} = \frac{\partial Q}{\partial y}, \text{ and } \frac{\partial P}{\partial y} = -\frac{\partial Q}{\partial x}$$

which immediately gives

$$\oint_C f(z)dz = 0$$

as required.

$\square$

Cauchy's theorem is so useful that pure mathematicians have spent much time and effort reducing the conditions for its validity. It is valid if $f(z)$ is analytic inside and not necessarily on C (as has already been said). This means that it is possible to extend it to regions that are semi-infinite (for example, a

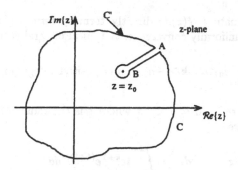

Figure 7.3: The indented contour

half plane). This is crucial for a text on Laplace Transforms. Let us look at another consequence of Cauchy's theorem.

Suppose $f(z)$ is analytic inside a closed curve C except at the point $z = z_0$ where it is singular. Then

$$\oint_C f(z)dz$$

may or may not be zero. If we *exclude* the point $z = z_0$ by drawing the contour C' (see Figure 7.3) where

$$\oint_{C'} = \int_C + \int_{AB} + \int_{BA} - \int_{\gamma}$$

and γ is a small circle surrounding $z = z_0$, then

$$\oint_{C'} f(z)dz = 0$$

as $f(z)$ is analytic inside C' by construction. Hence

$$\int_C f(z)dz = \int_{\gamma} f(z)dz$$

since

$$\int_{AB} f(z)dz + \int_{BA} f(z)dz = 0 \text{ as they are equal and opposite.}$$

In order to evaluate $\int_C f(z)dz$ therefore we need only to evaluate $\int_{\gamma} f(z)dz$ where γ is a small circle surrounding the singularity $(z = z_0)$ of $f(z)$. Now we apply the theory of Laurent series introduced in the last section. In order to do this with success, we restrict attention to the case where $z = z_0$ is a pole. Inside γ, $f(z)$ can be represented by the Laurent series

$$\frac{a_{-n}}{(z - z_0)^n} + \cdots + \frac{a_{-2}}{(z - z_0)^2} + \frac{a_{-1}}{(z - z_0)} + a_0 + \sum_{k=1}^{\infty} a_k(z - z_0)^k.$$

We can thus evaluate $\int_\gamma f(z)dz$ directly term by term. This is valid as the Laurent series is uniformly convergent. A typical integral is thus

$$\int_\gamma (z - z_0)^k dz \quad k \geq -n, \quad n \text{ any positive integer or zero.}$$

On γ, $z - z_0 = \epsilon e^{i\theta}$, $0 \leq \theta < 2\pi$ where ϵ is the radius of the circle γ and $dz = i\epsilon e^{i\theta} d\theta$. Hence

$$\int_\gamma (z - z_0)^k dz = \int_0^{2\pi} i\epsilon^{k+1} e^{(k+1)i\theta} d\theta$$

$$= \left[\frac{i\epsilon^{k+1}}{i(k+1)} e^{i(k+1)\theta} \right]_0^{2\pi} \quad \text{if } k \neq -1$$

$$= 0.$$

If $k = -1$

$$\int_\gamma (z - z_0)^{-1} dz = \int_0^{2\pi} \frac{i\epsilon e^{i\theta}}{\epsilon e^{i\theta}} d\theta = 2\pi i.$$

Thus

$$\int_\gamma (z - z_0)^k dz = \begin{cases} 0 & k \neq -1 \\ 2\pi i & k = -1. \end{cases}$$

Hence

$$\int_\gamma f(z)dz = 2\pi i a_{-1}$$

where a_{-1} is the coefficient of $\frac{1}{z-z_0}$ in the Laurent series of $f(z)$ about the singularity at $z = z_0$ called the *residue* of $f(z)$ at $z = z_0$. We thus, courtesy of Cauchy's theorem, arrive at the residue theorem.

Theorem 7.4 (Residue Theorem) *If $f(z)$ is analytic inside C except at points $z_1, z_2, \ldots, z_N$ where $f(z)$ has poles then*

$$\oint_C f(z)dz = 2\pi i \sum_{k=1}^{N} (\text{sum of residues of } f(z) \text{ at } z = z_k).$$

Proof This result follows immediately on a straightforward generalisation of the case just considered to the case of $f(z)$ possessing N poles inside C. No further elaboration is either necessary or given.

□

If $f(z)$ is a function whose only singularities are poles, it is called meromorphic. Meromorphic functions when integrated around closed contours C that contain at least one of the singularities of $f(z)$ lead to simple integrals. Distorting these closed contours into half or quarter planes leads to one of the most

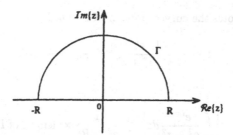

Figure 7.4: The semi-circular contour C

widespread of the applications of the residue theorem, namely the evaluation of real but improper integrals such as

$$\int_0^\infty \frac{dx}{1+x^4} \quad \text{or} \quad \int_0^\infty \frac{\cos(\pi x)}{a^2+x^2} dx.$$

These are easily evaluated by residue calculus (i.e. application of the residue theorem). The first by $\oint_C \frac{dz}{1+z^4}$ where C is a semi-circle in the upper half plane whose radius then tends to infinity and the second by considering $\oint_C \frac{e^{i\pi z}}{a^2+z^2} dz$ over a similar contour. We shall do these examples as illustrations, but point to books on complex variables for further examples. It is the evaluation of contour integrals that is a skill required for the interpretation of the Inverse Laplace Transform.

Example 7.5 *Use suitable contours C to evaluate the two real integrals*

(i) $\int_0^\infty \frac{\cos(\pi x)}{a^2+x^2} dx$

(ii) $\int_0^\infty \frac{dx}{1+x^4}.$

Solution

(i) For the first part, we choose the contour C shown in Figure 7.4, that is a semi-circular contour on the upper half plane. We consider the integral

$$\oint_C \frac{e^{i\pi z}}{a^2+z^2} dz.$$

Now,

$$\oint_C \frac{e^{i\pi z}}{a^2+z^2} dz = \int_\Gamma \frac{e^{i\pi z}}{a^2+z^2} dz + \int_{-R}^R \frac{e^{i\pi z}}{a^2+z^2} dz$$

where Γ denotes the curved portion of C. On Γ,

$$\left| \frac{e^{i\pi z}}{z^2 + a^2} \right| \leq \frac{1}{R^2 + a^2}$$

hence

$$\left| \int_\Gamma \frac{e^{i\pi z}}{a^2 + z^2} dz \right| \leq \frac{1}{a^2 + R^2} \times \text{ length of } \Gamma$$

$$= \frac{\pi R}{a^2 + R^2} \to 0 \text{ as } R \to \infty.$$

Hence, as $R \to \infty$

$$\oint_C \frac{e^{i\pi z}}{a^2 + z^2} dz \to \int_{-\infty}^{\infty} \frac{e^{i\pi z}}{a^2 + x^2} dx \text{ as on the real axis } z = x.$$

Using the residue theorem

$$\oint_C \frac{e^{i\pi z}}{a^2 + z^2} dz = 2\pi i \text{ \{residue at } z = ia\}.$$

The residue at $z = ia$ is given by the simple formula

$$\lim_{z \to z_0} [(z - z_0)f(z)] = a_{-1}$$

where the evaluation of the limit usually involves L'Hôpital's rule. Thus we have

$$\lim_{z \to ia} \frac{(z - ia)e^{\pi iz}}{z^2 + a^2} = \frac{e^{-\pi a}}{2ai}.$$

Thus, letting $R \to \infty$ gives

$$\int_{-\infty}^{\infty} \frac{e^{i\pi z}}{a^2 + x^2} dx = \frac{\pi}{a} e^{-\pi a}.$$

Finally, note that

$$\int_{-\infty}^{\infty} \frac{e^{i\pi z}}{a^2 + x^2} dx = \int_{-\infty}^{\infty} \frac{\cos(\pi x)}{a^2 + x^2} dx + i \int_{-\infty}^{\infty} \frac{\sin(\pi x)}{a^2 + x^2} dx$$

and

$$\int_{-\infty}^{\infty} \frac{\cos(\pi x)}{a^2 + x^2} dx = 2 \int_0^{\infty} \frac{\cos(\pi x)}{a^2 + x^2} dx.$$

Thus,

$$\int_0^{\infty} \frac{\cos(\pi x)}{a^2 + x^2} dx = \frac{\pi}{2a} e^{-\pi a}.$$

(ii) We could use a quarter circle ($x > 0$ and $y > 0$) for this problem as this only requires the calculation of a single residue $z = \frac{1+i}{\sqrt{2}}$. However, using the same contour (Figure 7.4) is in the end easier. The integral

$$\int_C \frac{dz}{1 + z^4} = 2\pi i \{\text{sum of residues}\}.$$

The residues are at the two solutions of $z^4 + 1 = 0$ that lie inside C, i.e. $z = \frac{1+i}{\sqrt{2}}$ and $z = \frac{-1+i}{\sqrt{2}}$. With $f(z) = \frac{1}{z^4}$, the two values of the residues calculated using the same formula as before are $\frac{-1-i}{4\sqrt{2}}$ and $\frac{1-i}{4\sqrt{2}}$. Their sum is $\frac{-i}{2\sqrt{2}}$. Hence

$$\int_C \frac{dz}{1 + z^4} = \frac{\pi}{\sqrt{2}}$$

on Γ (the curved part of C)

$$\left| \frac{1}{1 + z^4} \right| < \frac{1}{R^4 - 1}.$$

Hence

$$\left| \int_\Gamma \frac{dz}{1 + z^4} \right| < \frac{\pi R}{R^4 - 1} \to 0 \text{ as } R \to \infty.$$

Thus

$$\int_{-\infty}^{\infty} \frac{dx}{1 + x^4} = \frac{\pi}{\sqrt{2}}.$$

Hence

$$\int_0^{\infty} \frac{dx}{1 + x^4} = \frac{1}{2} \int_{-\infty}^{\infty} \frac{dx}{1 + x^4} = \frac{\pi}{2\sqrt{2}}.$$

This is all we shall do on this large topic here. We need to move on and consider branch points.

7.4   Branch Points

For the applications of complex variables required in this book, we need one further important development. Functions such as square roots are double valued for real variables. In complex variables, square roots and the like are called functions with branch points.

The function $w = \sqrt{z}$ has a branch point at $z = 0$. To see this and to get a feel for the implications, write $z = re^{i\theta}$ and as θ varies from 0 to 2π, z describes a circle radius r in the z plane, only a semi circle radius $\sqrt{r}$ is described in the w plane. In order for w to return to its original value, θ has to reach 4π. Therefore there are actually two planes superimposed upon one another in the z plane (θ varying from 0 to 4π), under the mapping each point on the w plane can arise from one point on (one of) the z planes. The two z planes are sometimes called Riemann sheets, and the place where one sheet ends and the next one starts is

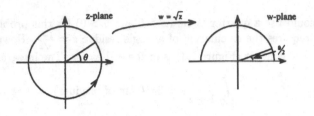

Figure 7.5: The mapping $w = \sqrt{z}$

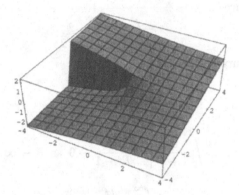

Figure 7.6: A 3D visualisation of $w = \sqrt{z}$. The horizontal axes are the Argand diagram and the quantity $\Im\sqrt{z}$ is displayed on the vertical axis. The cut along the negative real axis is clearly visible

marked by a cut from $z = 0$ to $z = \infty$ and is typically the positive real axis (see Figure 7.5). Perhaps a better visualisation is given in Figure 7.6 in which the vertical axis is the imaginary part of $\sqrt{z}$. It is clearly seen that a discontinuity develops along the negative real axis, hence the cut. The function $w = \sqrt{z}$ has two sheets, whereas the function $w = z^{\frac{1}{N}}$ (N a positive integer) has N sheets and the function $w = \ln(z)$ has infinitely many sheets.

When a contour is defined for the purposes of integration, it is not permitted for the contour to cross a cut. Figure 7.7 shows a cut and a typical way in which crossing it is avoided. In this contour, a complete circuit of the origin is desired but rendered impossible because of the cut along the positive real axis. So we start at B just above the real axis, go around the circle as far as C below B. Then along CD just below the real axis, around a small circle surrounding the origin as far as A then along and just above the real axis to complete the circuit at B. In order for this contour (called a *key hole* contour) to approach the desired circular contour, $|AD| \to 0$, $|BC| \to 0$ and the radius of the small circle surrounding the origin also tends to zero. In order to see this process in operation let us do an example.

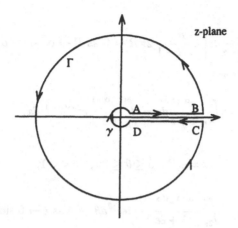

Figure 7.7: The keyhole contour

Example 7.6 *Find the value of the real integral*

$$\int_0^\infty \frac{x^{\alpha-1}}{1+x}dx \text{ where } 0 < \alpha < 1 \text{ } \alpha \text{ a real constant.}$$

Solution In order to find this integral, we evaluate the complex contour integral

$$\int_C \frac{z^{\alpha-1}}{1+z}dz$$

where C is the keyhole contour shown in Figure 7.7.

Sometimes there is trouble because there is a singularity where the cut is usually drawn. We shall meet this in the next example, but the positive real axis is free of singularities in this example. The simple pole at $z = -1$ is inside C and the residue is given by

$$\lim_{z\to-1}(z+1)\frac{z^{\alpha-1}}{(z+1)} = e^{(\alpha-1)i\pi}.$$

Thus

$$\int_C \frac{z^{\alpha-1}}{1+z}dz = \int_\Gamma \frac{z^{\alpha-1}}{1+z}dz + \int_{CD} \frac{z^{\alpha-1}}{1+z}dz + \int_\gamma \frac{z^{\alpha-1}}{1+z}dz + \int_{AB} \frac{z^{\alpha-1}}{1+z}dz$$

$$= 2\pi i e^{(\alpha-1)i\pi}.$$

Each integral is taken in turn.

On Γ

$$z = Re^{i\theta}, \quad 0 \le \theta \le 2\pi, \quad R \gg 1,$$

$$\left|\frac{z^{\alpha-1}}{1+z}\right| < \frac{R^{\alpha-1}}{R-1};$$

hence

$$\left| \int_C \frac{z^{\alpha-1}}{1+z} dz \right| < \frac{R^{\alpha-1}}{R-1} 2\pi R \to 0 \text{ as } R \to \infty \text{ since } \alpha < 1.$$

On CD, $z = xe^{2\pi i}$ so

$$\int_{CD} \frac{z^{\alpha-1}}{1+z} dz = \int_R^\epsilon \frac{x^{\alpha-1}}{1+x} e^{2\pi i(\alpha-1)} dx.$$

On γ

$$z = \epsilon e^{i\theta}, \quad 0 \le \theta \le 2\pi, \quad \epsilon << 1,$$

so

$$\int_\gamma \frac{z^{\alpha-1}}{1+z} dz = \int_{2\pi}^0 \frac{\epsilon^{\alpha-1} e^{i(\alpha-1)\theta}}{1+\epsilon e^{i\theta}} i\epsilon e^{i\theta} d\theta \to 0 \text{ as } \epsilon \to 0 \text{ since } \alpha > 0.$$

Finally, on AB, $z = x$ so

$$\int_{AB} \frac{z^{\alpha-1}}{1+z} dz = \int_\epsilon^R \frac{x^{\alpha-1}}{1+x} dx.$$

Adding all four integrals gives, letting $\alpha \to 0$ and $R \to \infty$,

$$\int_\infty^0 \frac{x^{\alpha-1}}{1+x} e^{2\pi i(\alpha-1)} dx + \int_0^\infty \frac{x^{\alpha-1}}{1+x} dx = 2\pi i e^{(\alpha-1)i\pi}.$$

Rearranging gives

$$\int_0^\infty \frac{x^{\alpha-1}}{1+x} dx = \frac{\pi}{\sin(\alpha\pi)}.$$

This is only valid if $0 < \alpha < 1$, the integral is singular otherwise.

To gain experience with a different type of contour, here is a second example. After this, we shall be ready to evaluate the so called Bromwich contour for the Inverse Laplace Transform.

Example 7.7 *Use a semi-circular contour in the upper half plane to evaluate*

$$\int_C \frac{\ln z}{z^2 + a^2} dz \quad (a > 0) \text{ real and positive}$$

and deduce the values of two real integrals.

Solution Figure 7.8 shows the semi-circular contour. It is indented at the origin as $z = 0$ is an essential singularity of the integrand $\dfrac{\ln z}{z^2 + a^2}$. Thus

$$\int_C \frac{\ln z}{z^2 + a^2} dz = 2\pi i \{\text{Residue at } z = ia\}$$

provided R is large enough and the radius of the small semi-circle $\gamma \to 0$.

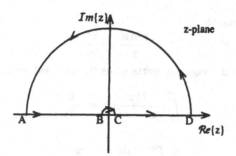

Figure 7.8: The indented semi-circular contour C.

The residue at $z = ia$ is given by

$$\frac{\ln(ia)}{2ia} = \frac{\pi}{4a} - i\frac{\ln(a)}{2a} \quad (a > 0)$$

so the right-hand side of the residue theorem becomes

$$\frac{\pi \ln(a)}{a} + i\frac{\pi^2}{2a}.$$

On the semi-circular indent γ,

$$\int_\gamma \frac{\ln z}{z^2 + a^2} dz = \int_\pi^0 \frac{\ln(\epsilon e^{i\theta})i\epsilon e^{i\theta}}{\epsilon^2 e^{2i\theta} + a^2} d\theta \to 0 \text{ as } \epsilon \to 0.$$

(This is because $\epsilon \ln \epsilon \to 0$ as $\epsilon \to 0$.) Thus

$$\int_C \frac{\ln z}{z^2 + a^2} dz = \frac{\pi \ln a}{a} + i\frac{\pi^2}{2a}.$$

Now, as we can see from Figure 7.8,

$$\int_C \frac{\ln z}{z^2 + a^2} dz = \int_\Gamma \frac{\ln z}{z^2 + a^2} dz + \int_{AB} \frac{\ln z}{z^2 + a^2} dz + \int_\gamma \frac{\ln z}{z^2 + a^2} dz + \int_{CD} \frac{\ln z}{z^2 + a^2} dz.$$

On Γ, $z = Re^{i\theta}$, $0 \le \theta \le \pi$, and

$$\left| \frac{\ln z}{z^2 + a^2} \right| \le \frac{\ln R}{R^2 - a^2}$$

so

$$\left| \int_\Gamma \frac{\ln z}{z^2 + a^2} dz \right| \le \frac{\pi R \ln R}{R^2 - a^2} \to 0 \text{ as } R \to \infty.$$

Thus letting $R \to \infty$, the radius of $\gamma \to 0$ and evaluating the straight line integrals via $z = x$ on CD and $z = xe^{i\pi}$ on AB gives

$$\int_C \frac{\ln z}{z^2 + a^2} dz = \int_\infty^0 \frac{\ln xe^{i\pi}}{x^2 + a^2} e^{i\pi} dx + \int_0^\infty \frac{\ln x}{x^2 + a^2} dx$$

$$= 2 \int_0^\infty \frac{\ln x}{x^2 + a^2} dx + i\pi \int_0^\infty \frac{dx}{x^2 + a^2}.$$

The integral is equal to

$$\frac{\pi \ln a}{a} + i\frac{\pi^2}{2a}$$

so equating real and imaginary parts gives the two real integrals

$$\int_0^\infty \frac{\ln x}{x^2 + a^2} dx = \frac{\pi \ln a}{2a}$$

and

$$\int_0^\infty \frac{dx}{x^2 + a^2} = \frac{\pi}{2a}.$$

The second integral is an easily evaluated arctan standard form.

7.5 The Inverse Laplace Transform

We are now ready to derive and use the formula for the inverse Laplace Transform. It is a surprise to engineers that the inverse of a transform so embedded in real variables as the Laplace Transform requires so deep a knowledge of complex variables for its evaluation. It should not be so surprising having studied the last two chapters. We state the inverse transform as a theorem.

Theorem 7.8 *If the Laplace Transform of $F(t)$ exists, that is $F(t)$ is of exponential order and*

$$f(s) = \int_0^\infty e^{-st} F(t) dt$$

then

$$F(t) = \lim_{k \to \infty} \left\{ \frac{1}{2\pi i} \int_{\sigma - ik}^{\sigma + ik} f(s) e^{st} ds \right\} \quad t > 0$$

where $|F(t)| \leq e^{Mt}$ for some positive real number M and σ is another real number such that $\sigma > M$.

Proof The proof of this has already been outlined in Section 6.2 of the last chapter. However, we have now done enough formal complex variable theory to give a more complete proof. The outline remains the same in that we define $F_k(\omega)$ as in the last chapter, namely

$$F_k(\omega) = \int_0^\infty e^{-(k+i\omega)x} f(x) dx$$

and rewrite this in notation more suited to Laplace Transforms, i.e. x becomes t, $k + i\omega$ becomes s and the functions are renamed. $f(x)$ becomes $F(t)$ and $F_k(\omega)$ becomes $f(s)$. However, the mechanics of the proof follows as before with Equation 6.2. Using the new notation, these two equations convert to

$$f(s) = \int_0^\infty e^{-st} F(t) dt$$

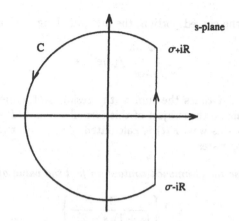

Figure 7.9: The Bromwich contour

and

$$\frac{1}{2\pi}\int_{-\infty}^{\infty} e^{st} f(s) d\{\Im(s)\} = \begin{cases} 0 & t < 0 \\ F(t) & t > 0 \end{cases}$$

where the integral on the left is a real integral. $d\{\Im(s)\} = d\omega$ a real differential. Converting this to a complex valued integral, formally done by recognising that $ds = id\{\Im(s)\}$, gives the required formula, viz.

$$F(t) = \frac{1}{2\pi}\int_{s_0}^{s_1} e^{st} f(s) ds$$

where s_0 and s_1 represent the infinite limits ($k - i\infty$, $k + i\infty$ in the notation of Chapter 6). Now, however we can be more precise. The required behaviour of $F(t)$ means that the real part of s must be at least as large as σ, otherwise $|F(t)|$ does not $\to 0$ as $t \to \infty$ on the straight line between s_0 and s_1. This line is parallel to the imaginary axis. The theorem is thus formally established.

□

The way of evaluating this integral is via a closed contour of the type shown in Figure 7.9. This contour, often called the Bromwich contour, consists of a portion of a circle, radius R, together with a straight line segment connecting the two points $\sigma - iR$ and $\sigma + iR$. The real number σ must be selected so that all the singularities of the function $f(s)$ are to the left of this line. This follows from the conditions of Theorem 7.8. The integral

$$\int_C f(s) e^{st} ds$$

where C is the Bromwich contour is evaluated using Cauchy's residue theorem, perhaps with the addition of one or two cuts. The integral itself is the sum of the integral over the curved portion, the integral along any cuts present and

$$\int_{\sigma-iR}^{\sigma+iR} f(s)e^{st}ds$$

and the whole is $2\pi i$ times the sum of the residues of $f(s)e^{st}$ inside C. The above integral is made the subject of this formula, and as $R \to \infty$ this integral becomes $F(t)$. In this way, $F(t)$ is calculated. Let us do two examples to see how the process operates.

Example 7.9 *Use the Bromwich contour to find the value of*

$$\mathcal{L}^{-1}\left\{\frac{1}{(s+1)(s-2)^2}\right\}.$$

Solution It is quite easy to find this particular inverse Laplace Transform using partial fractions as in Chapter 2; however it serves as an illustration of the use of the contour integral method. In the next example, there are no alternative direct methods.

Now,

$$\int_C \frac{e^{st}}{(s+1)(s-2)^2}ds = 2\pi i\{\text{sum of residues}\}$$

where C is the Bromwich contour of Figure 7.9. The residue at $s = 1$ is given by

$$\lim_{s \to -1}(s+1)\frac{e^{st}}{(s+1)(s-2)^2} = \frac{1}{9}e^{-t}.$$

The residue at $s = 2$ is given by

$$\lim_{s \to 2}\frac{d}{ds}\frac{e^{st}}{(s+1)} = \left[\frac{(s+1)te^{st} - e^{st}}{(s+1)^2}\right]_{s=2} = \frac{1}{9}(3te^{2t} - e^{2t}).$$

Thus

$$\int_C \frac{e^{st}}{(s+1)(s-2)^2}ds = 2\pi i\left\{\frac{1}{9}(e^{-t} + 3te^{2t} - e^{2t})\right\}.$$

Now,

$$\int_C \frac{e^{st}}{(s+1)(s-2)^2}ds = \int_\Gamma \frac{e^{st}}{(s+1)(s-2)^2}ds + \int_{\sigma-iR}^{\sigma+iR} \frac{e^{st}}{(s+1)(s-2)^2}ds$$

and the first integral $\to 0$ as the radius of the Bromwich contour $\to \infty$. Thus

$$\mathcal{L}^{-1}\left\{\frac{1}{(s+1)(s-2)^2}\right\} = \frac{1}{2\pi i}\int_{\sigma-i\infty}^{\sigma+i\infty} \frac{e^{st}}{(s+1)(s-2)^2}ds = \frac{1}{9}(e^{-t} + 3te^{2t} - e^{2t}).$$

A result easily confirmed by the use of partial fractions.

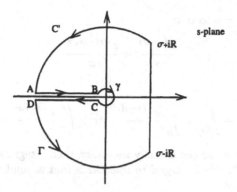

Figure 7.10: The cut Bromwich contour

Example 7.10 *Find*

$$\mathcal{L}^{-1}\left\{\frac{e^{-a\sqrt{s}}}{s}\right\}$$

where a is a real constant.

Solution The presence of $e^{-a\sqrt{s}}$ means that the origin $s = 0$ is a branch point. It is thus necessary to use the cut Bromwich contour C' as shown in Figure 7.10. Now,

$$\int_{C'}\frac{e^{st-a\sqrt{s}}}{s}ds = 0$$

by Cauchy's theorem as there are no singularities of the integrand inside C'. We now evaluate the contour integral by splitting it into its five parts:-

$$\int_{C'}\frac{e^{st-a\sqrt{s}}}{s}ds = \int_{\Gamma} + \int_{\sigma-iR}^{\sigma+iR} + \int_{AB} + \int_{\gamma} + \int_{CD} = 0$$

and consider each bit in turn. The radius of the large circle Γ is R and the radius of the small circle γ is ϵ. On Γ, $s = Re^{i\theta} = R\cos(\theta) + iR\sin(\theta)$ and on the left hand side of the s-plane $\cos\theta < 0$. This means that, on Γ,

$$\left|\frac{e^{st-a\sqrt{s}}}{s}\right| = \left|\frac{e^{Rt\cos(\theta)-a\sqrt{R}\cos(\frac{1}{2}\theta)}}{R}\right|$$

and by the estimation lemma,

$$\int_{\Gamma} \to 0 \text{ as } R \to \infty.$$

The second integral is the one required and we turn our attention to the third integral along AB. On AB, $s = xe^{i\pi}$ whence

$$\frac{e^{st-a\sqrt{s}}}{s} = -\frac{e^{-xt-ai\sqrt{x}}}{x}$$

whereas on CD $s = x$ so that

$$\frac{e^{st-a\sqrt{s}}}{s} = \frac{e^{-xt+ai\sqrt{x}}}{x}.$$

This means that

$$\int_{AB} + \int_{CD} = \int_R^\epsilon \frac{e^{-xt-ai\sqrt{x}}}{x} dx + \int_\epsilon^R \frac{e^{-xt+ai\sqrt{x}}}{x} dx.$$

It is the case that if the cut is really necessary, then integrals on either side of it never cancel. The final integral to consider is that around γ. On γ $s = \epsilon e^{i\theta}$, $-\pi \le \theta < \pi$, so that

$$\int_\gamma \frac{e^{-st-a\sqrt{s}}}{s} ds = \int_\pi^{-\pi} \frac{e^{-\epsilon e^{i\theta}t-a\sqrt{\epsilon e^{\frac{1}{2}i\theta}}}}{\epsilon e^{i\theta}} i\epsilon e^{i\theta} d\theta.$$

Now, as $\epsilon \to 0$

$$\int_\gamma \to \int_\pi^{-\pi} i d\theta = -2\pi i.$$

Hence, letting $R \to \infty$ and $\epsilon \to 0$ gives

$$\frac{1}{2\pi i} \int_{\sigma-i\infty}^{\sigma+i\infty} \frac{e^{st-a\sqrt{s}}}{s} ds = 1 - \frac{1}{2\pi i} \left\{ \int_\infty^0 \frac{e^{-xt-ai\sqrt{x}}}{x} dx + \int_0^\infty \frac{e^{-xt+ai\sqrt{x}}}{x} dx \right\}$$

$$= 1 - \frac{1}{\pi} \int_0^\infty \frac{e^{-xt} \sin(a\sqrt{x})}{x} dx.$$

As the left hand side is $F(t)$, the required inverse is

$$1 - \frac{1}{\pi} \int_0^\infty \frac{e^{-xt} \sin(a\sqrt{x})}{x} dx.$$

This integral can be simplified (if that is the correct word) by the substitution $x = u^2$ then using differentiation under the integral sign. Omitting the details, it is found that

$$\frac{1}{\pi} \int_0^\infty \frac{e^{-xt} \sin(a\sqrt{x})}{x} dx = \operatorname{erf}\left(\frac{a}{2\sqrt{t}}\right)$$

where erf is the error function (see Chapter 3). Hence

$$\mathcal{L}^{-1}\left\{ \frac{e^{-a\sqrt{s}}}{s} \right\} = 1 - \operatorname{erf}\left(\frac{a}{2\sqrt{t}}\right) = \operatorname{erfc}\left(\frac{a}{2\sqrt{t}}\right)$$

where

$$\operatorname{erfc}(p) = \frac{2}{\sqrt{\pi}} \int_p^\infty e^{-t^2} dt$$

is the complementary error function.

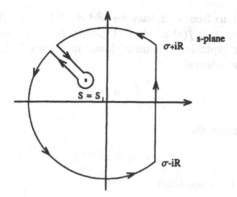

Figure 7.11: The contour C distorted around the branch point at $s = s_1$

7.6 Using the Inversion Formula in Asymptotics

We saw in Chapter 5 how it is possible to use asymptotic expansions in order to gain information from a differential equation even if it was not possible to solve it in closed form. Let us return to consider this kind of problem, now armed with the complex inversion formula. It is often possible to approximate the inversion integral using asymptotic analysis when exact inversion cannot be done. Although numerical techniques are of course also applicable, asymptotic methods are often more appropriate. After all, there is little point employing numerical methods at this late stage when it would have probably been easier to use numerical techniques from the outset on the original problem! Of course, by so doing all the insight gained by adopting analytical methods would have been lost. Not wishing to lose this insight, we press on with asymptotics. The following theorem embodies a particularly useful asymptotic property for functions that have branch points.

Theorem 7.11 *If $\bar{f}(s)$ is $O(1/|s|)$ as $s \to \infty$, the singularity with largest real part is at $s = s_1$, and in some neighbourhood of s_1*

$$\bar{f}(s) = (s - s_1)^k \sum_{n=0}^{\infty} a_n (s - s_1)^n \quad -1 < k < 0,$$

then

$$f(t) = -\frac{1}{\pi} e^{s_1 t} \sin(k\pi) \sum_{n=0}^{\infty} a_n (-1)^n \frac{\Gamma(n + k + 1)}{t^{n+k+1}} \quad \text{as } t \to \infty.$$

Proof In this proof we shall assume that $s = s_1$ is the only singularity and that it is a branch point. This is acceptable as we are seeking to prove an asymptotic result and as long as the singularities of $\bar{f}(s)$ are isolated only the one with the largest real part is of interest. If there is a tie, each singularity is considered and the contributions from each may be added. When the Bromwich contour is used, and the function $\bar{f}(s)$ has branch points, it has to be distorted around these points and a typical distortion is shown in Figure 7.11 and the object is to approximate the integral

$$\frac{1}{2\pi i}\int_C e^{st}\bar{f}(s)ds.$$

On the lower Riemann sheet

$$s = s_1 + xe^{-i\pi}$$

and on the upper Riemann sheet

$$s = s_1 + xe^{i\pi}.$$

The method used to evaluate the integral is to invoke Watson's lemma as follows

$$f(t) = \frac{1}{2\pi i}\int_\infty^0 e^{(s_1-x)t}\bar{f}(s_1+xe^{-i\pi})e^{-i\pi}dx + \frac{1}{2\pi i}\int_0^\infty e^{(s_1-x)t}\bar{f}(s_1+xe^{i\pi})e^{i\pi}dx.$$

These integrals combine to give

$$f(t) = -\frac{1}{2\pi i}\int_0^\infty e^{(s_1-x)t}[\bar{f}(s_1+xe^{i\pi}) - \bar{f}(s_1+xe^{-i\pi})]dx.$$

Watson's lemma is now used to expand the two functions in the square bracket in powers of x. It really is just algebra, so only the essential steps are given.

Assume that $\bar{f}(s)$ can be expressed in the manner indicated in the theorem so

$$\bar{f}(s_1+xe^{i\pi}) = (xe^{i\pi})^k \sum_{n=0}^\infty a_n(xe^{i\pi})^n$$

and

$$\bar{f}(s_1+xe^{-i\pi}) = (xe^{-i\pi})^k \sum_{n=0}^\infty a_n(xe^{-i\pi})^n.$$

These two expressions are subtracted so that as is typical, it is the $e^{\pm ik\pi}$ terms that prevent complete cancellation. Thus

$$\bar{f}(s_1+xe^{i\pi}) - \bar{f}(s_1+xe^{-i\pi}) = x^k 2i\sin(k\pi)\sum_{n=0}^\infty a_n(-1)^n x^n.$$

We are now faced with the actual integration, so

$$\frac{1}{2\pi i}\int_0^\infty e^{(s_1-x)t}[\bar{f}(s_1+xe^{i\pi}) - \bar{f}(s_1+xe^{-i\pi})]dx$$

$$= \frac{\sin(k\pi)}{\pi}\int_0^\infty \sum_{n=0}^\infty a_n(-1)^n x^{n+k} e^{(s_1-x)t}dx$$

$$= \frac{\sin(k\pi)}{\pi} e^{s_1 t} \sum_{n=0}^{\infty} a_n (-1)^n \int_0^\infty x^{n+k} e^{-xt} dx$$

provided exchanging summation and integral signs is valid. Now the integral definition of the gamma function

$$\Gamma(n) = \int_0^\infty u^{n-1} e^{-u} du$$

comes to our aid to evaluate the integral that remains. Using the substitution $u = xt$, we have

$$\int_0^\infty x^{n+k} e^{-xt} dx = \frac{1}{t^{n+k+1}} \Gamma\{n+k+1\}.$$

This then formally establishes the theorem, leaving only one or two niggly questions over the legality of some of the steps. One common query is how is it justifiable to integrate with respect to x as far as infinity when we are supposed to be "close to s_1" the branch point of $\bar{f}(s)$? It has to be remembered that this is an approximation valid for large t, hence the major contribution will come from that part of the contour that is near to $x = 0$. The presence of infinity in the limit of the integral is therefore somewhat misleading. (For those familiar with boundary layers in fluid mechanics, it is not uncommon for "infinity" to be as small as 0.2!)

□

If the singularity at $s = s_1$ is a pole, then the integral can be evaluated directly by use of the residue theorem, giving

$$f(t) = \frac{1}{2\pi i} \int_C e^{st} \bar{f}(s) ds$$

$$= R e^{s_1 t}$$

where R is the residue of $\bar{f}(s)$ at $s = s_1$. The following example illustrates the use of this theorem to approximate the behaviour of the solution to a BVP for large y.

Example 7.12 *Find the asymptotic behaviour as $y \to \infty$ for fixed x of the solution of the partial differential equation*

$$\frac{\partial^2 \phi}{\partial y^2} = \frac{\partial^2 \phi}{\partial x^2} - \phi, \quad x > 0, \quad y > 0,$$

such that

$$\phi(x,0) = \frac{\partial \phi}{\partial y}(x,0) = 0, \quad \phi(0,y) = 1.$$

Solution Taking the Laplace Transform of the given equation with respect to y using the by now familiar notation leads to the following ODE for $\bar\phi$:

$$(1+s^2)\bar\phi = \frac{d^2\bar\phi}{dx^2}$$

where the boundary conditions at $y=0$ have already been utilised. At $x=0$ we require that $\phi = 1$ for all y. This transforms to $\bar\phi = 1/s$ at $x=0$ which thus gives the solution

$$\bar\phi(x,s) = \frac{e^{-x\sqrt{1+s^2}}}{s}.$$

We have discarded the part of the complementary function $e^{x\sqrt{1+s^2}}$ as it does not tend to zero for large s and so cannot represent a function of s that can have arisen from a Laplace Transform (see Chapter 2). The problem would be completely solved if we could invert the Laplace Transform

$$\bar u = \frac{1}{s}e^{-x\sqrt{1+s^2}}$$

but alas this is not possible in closed form. The best we can do is to use Theorem 7.11 to find an approximation valid for large values of y. Now, $\bar u$ has a simple pole at $s=0$ and two branch points at $s=\pm i$. As the real part of all of these singularities is the same, viz. zero, all three contribute to the leading term for large y. At $s=0$ the residue is straightforwardly e^{-x}. Near $s=i$ the expansion

$$\bar\phi(x,s) = \frac{1}{i}[1 - (2i)^{1/2}(s-i)^{1/2}x + \cdots]$$

is valid. Near $s=-i$ the expansion

$$\bar\phi(x,s) = -\frac{1}{i}[1 - (-2i)^{1/2}(s+i)^{1/2}x + \cdots]$$

is valid. The value of $\phi(x,y)$, which is precisely the integral

$$\phi(x,y) = \frac{1}{2\pi i}\int_C e^{sy}\bar\phi(x,s)ds$$

is approximated by the following three terms; e^{-x} from the pole at $s=0$, and the two contribution from the two branch points at $s=\pm i$. Using Theorem 7.11 these are

$$u \sim \frac{1}{\pi}e^{iy}\sin\left(\frac{1}{2}\pi\right)\left((2i)^{1/2}\frac{\Gamma\{3/2\}}{y^{3/2}}\right) \quad \text{near } s=i$$

and

$$u \sim \frac{1}{\pi}e^{-iy}\sin\left(-\frac{1}{2}\pi\right)\left((-2i)^{1/2}\frac{\Gamma\{3/2\}}{y^{3/2}}\right) \quad \text{near } s=-i.$$

The sum of all these dominant terms is

$$u \sim e^{-x} + \frac{2^{1/2}x}{\pi^{1/2}y^{3/2}}\cos\left(y+\frac{\pi}{4}\right).$$

This is the behaviour of the solution $\phi(x, y)$ for large values of y.

Further examples of the use of asymptotics can be found in specialist texts on partial differential equations, e.g. Williams (1980), Weinberger (1965). For more about asymptotic expansions, especially the rigorous side, there is nothing to better the classic text of Copson (1967).

7.7 Exercises

1. The following functions all have simple poles. Find their location and determine the residue at each pole.

   (i) $\dfrac{1}{1+z}$,

   (ii) $\dfrac{2z+1}{z^2 - z - 2}$

   (iii) $\dfrac{3z^2 + 2}{(z-1)(z^2 + 9)}$

   (iv) $\dfrac{3 + 4z}{z^3 + 3z^2 + 2z}$

   (v) $\dfrac{\cos(z)}{z}$

   (vi) $\dfrac{z}{\sin(z)}$

   (vii) $\dfrac{e^z}{\sin(z)}$.

2. The following functions all have poles. Determine their location, order, and all the residues.

   (i) $\left(\dfrac{z+1}{z-1}\right)^2$

   (ii) $\dfrac{1}{(z^2 + 1)^2}$

   (iii) $\dfrac{\cos(z)}{z^3}$

   (iv) $\dfrac{e^z}{\sin^2(z)}$.

3. Use the residue theorem to evaluate the following integrals:

   (i)
   $$\int_C \frac{2z}{(z-1)(z+2)(z+i)} dz$$
   where C is any contour that includes within it the points $z = 1$, $z = 2$ and $z = -i$.

   (ii)
   $$\int_C \frac{z^4}{(z-1)^3} dz$$
   where C is any contour that encloses the point $z = 1$.

   (iii)
   $$\int_0^\infty \frac{1}{x^6 + 1} dx.$$

   (iv)
   $$\int_0^\infty \frac{\cos(2\pi x)}{x^4 + x^2 + 1} dx.$$

4. Use the indented semi-circular contour of Figure 7.8 to evaluate the three real integrals:

(i) $\displaystyle\int_0^\infty \frac{(\ln x)^2}{x^4+1}dx$, (ii) $\displaystyle\int_0^\infty \frac{\ln x}{x^4+1}dx$, (iii) $\displaystyle\int_0^\infty \frac{x^\lambda}{x^2+1}dx$, $-1 < \lambda < 1$.

5. Determine the following inverse Laplace Transforms:

$$\text{(i) } \mathcal{L}^{-1}\left\{\frac{1}{s\sqrt{s+1}}\right\}, \text{(ii) } \mathcal{L}^{-1}\left\{\frac{1}{1+\sqrt{s+1}}\right\},$$

$$\text{(iii) } \mathcal{L}^{-1}\left\{\frac{1}{\sqrt{s}+1}\right\}, \text{(iv) } \mathcal{L}^{-1}\left\{\frac{1}{\sqrt{s}-1}\right\}.$$

You may find that the integrals

$$\int_0^\infty \frac{e^{-xt}}{(x+1)\sqrt{x}}dx = -\pi e^t[-1+\mathrm{erf}\sqrt{t}]$$

and

$$\int_0^\infty \frac{e^{-xt}\sqrt{x}}{(x+1)}dx = \sqrt{\frac{\pi}{t}} - \pi e^t \mathrm{erfc}\sqrt{t}$$

help the algebra!

6. Define the function $\varphi(t)$ via the inverse Laplace Transform

$$\varphi(t) = \mathcal{L}^{-1}\left\{\mathrm{erf}\frac{1}{s}\right\}.$$

Show that

$$\mathcal{L}\{\varphi(t)\} = \frac{2}{\sqrt{\pi s}} \sin\left(\frac{1}{\sqrt{s}}\right).$$

7. The zero order Bessel function can be defined by the series

$$J_0(xt) = \sum_{k=0}^\infty \frac{(-1)^k (\frac{1}{2}x)^{2k} t^{2k}}{(k!)^2}.$$

Show that

$$\mathcal{L}\{J_0(xt)\} = \frac{1}{s}\left(1+\frac{x^2}{s^2}\right)^{-1/2}.$$

8. Determine

$$\mathcal{L}^{-1}\left\{\frac{\cosh(x\sqrt{s})}{s\cosh(\sqrt{s})}\right\}$$

by direct use of the Bromwich contour.

9. Use the Bromwich contour to show that

$$\mathcal{L}^{-1}\{e^{-s^{1/3}}\} = \frac{3}{\pi}\int_0^\infty u^2 e^{-tu^3 - \frac{1}{2}u} \sin\left(\frac{u\sqrt{3}}{2}\right) du.$$

10. The function $\phi(x, t)$ satisfies the partial differential equation

$$\frac{\partial^2 \phi}{\partial x^2} - \frac{\partial^2 \phi}{\partial t^2} + \phi = 0$$

with

$$\phi(x, 0) = \frac{\partial \phi}{\partial t}(x, 0) = 0, \quad \phi(0, t) = t.$$

Use the asymptotic method of Section 7.6 to show that

$$\phi(x, t) \sim \frac{xe^t}{\sqrt{2\pi t^3}} \quad \text{as } t \to \infty$$

for fixed x.

Exercises 1.4

1. (a) $\ln t$ is singular at $t = 0$, hence the Laplace Transform does not exist.
   (b)
   $$\mathcal{L}\{e^{3t}\} = \int_0^\infty e^{3t} e^{-st} dt = \left[\frac{1}{3-s} e^{(3-s)t}\right]_0^\infty = \frac{1}{s-3}.$$
   (c) $e^{t^2} > |e^{Mt}|$ for any M for large enough t, hence the Laplace Transform does not exist (not of exponential order).
   (d) the Laplace Transform does not exist (singular at $t = 0$).
   (e) the Laplace Transform does not exist (singular at $t = 0$).
   (f) does not exist (infinite number of (finite) jumps), also not defined unless t is an integer.

2. Using the definition of Laplace Transform in each case, the integration is reasonably straightforward:

   $$\text{(a)} \quad \int_0^\infty e^{kt} e^{-st} dt = \frac{1}{s-k}$$

   as in part (b) of the previous question.
   (b) Integrating by parts gives,

   $$\mathcal{L}\{t^2\} = \int_0^\infty t^2 e^{-st} dt = \left[-\frac{t^2}{s} e^{-st}\right]_0^\infty + \int_0^\infty \frac{2t}{s} e^{-st} dt = \frac{2}{s} \int_0^\infty t e^{-st} dt.$$

   Integrating by parts again gives the result $\dfrac{2}{s^3}$.

(c) Using the definition of $\cosh(t)$ gives

$$\mathcal{L}\{\cosh(t)\} = \frac{1}{2}\left\{\int_0^\infty e^t e^{-st}\,dt + \int_0^\infty e^{-t}e^{-st}\,dt\right\}$$

$$= \frac{1}{2}\left\{\frac{1}{s-1} + \frac{1}{s+1}\right\} = \frac{s}{s^2-1}.$$

3. (a) This demands use of the first shift theorem, Theorem 1.3, which with $b = 3$ is

$$\mathcal{L}\{e^{-3t}F(t)\} = f(s+3)$$

and with $F(t) = t^2$, using part (b) of the last question gives the answer

$$\frac{2}{(s+3)^3}.$$

(b) For this part, we use Theorem 1.1 (linearity) from which the answer

$$\frac{4}{s^2} + \frac{6}{s-4}$$

follows at once.

(c) The first shift theorem with $b = 4$ and $F(t) = \sin(5t)$ gives

$$\frac{5}{(s+4)^2 + 25} = \frac{5}{s^2 + 8s + 41}.$$

4. When functions are defined in a piecewise fashion, the definition integral for the Laplace Transform is used and evaluated directly. For this problem we get

$$\mathcal{L}\{F(t)\} = \int_0^\infty e^{-st}F(t)\,dt = \int_0^1 te^{-st}\,dt + \int_1^2 (2-t)e^{-st}\,dt$$

which after integration by parts gives

$$\frac{1}{s^2}(1 - e^{-s})^2.$$

5. Using Theorem 1.8 we get

(a) $\displaystyle \mathcal{L}\{te^{2t}\} = \frac{d}{ds}\frac{1}{(s-2)} = \frac{1}{(s-2)^2}$

(b) $\displaystyle \mathcal{L}\{t\cos(t)\} = \frac{d}{ds}\frac{s}{1+s^2} = \frac{1-s^2}{(1+s^2)^2}$

The last part demands differentiating twice,

(c) $\displaystyle \mathcal{L}\{t^2\cos(t)\} = \frac{d^2}{ds^2}\frac{s}{1+s^2} = \frac{2s^3 - 6s}{(1+s^2)^3}.$

6. These two examples are not difficult: the first has application to oscillating systems and is evaluated directly, the second needs the first shift theorem with $b = 5$.

$$\text{(a)} \quad \mathcal{L}\{\sin(\omega t + \phi)\} = \int_0^\infty e^{-st} \sin(\omega t + \phi)dt$$

and this integral is evaluated by integrating by parts twice using the following trick. Let

$$I = \int_0^\infty e^{-st} \sin(\omega t + \phi)dt$$

then derive the formula

$$I = \left[-\frac{1}{s}e^{-st}\sin(\omega t + \phi) - \frac{\omega}{s^2}e^{-st}\cos(\omega t + \phi) \right]_0^\infty - \frac{\omega^2}{s^2}I$$

from which

$$I = \frac{s\sin(\phi) + \omega\cos(\phi)}{s^2 + \omega^2}.$$

$$\text{(b)} \quad \mathcal{L}\{e^{5t}\cosh(6t)\} = \frac{s-5}{(s-5)^2 - 36} = \frac{s-5}{s^2 - 10s - 11}.$$

7. This problem illustrates the difficulty in deriving a linear translation plus scaling property for Laplace Transforms. The zero in the bottom limit is the culprit. Direct integration yields:

$$\mathcal{L}\{G(t)\} = \int_0^\infty e^{-st}G(t)dt = \int_{-b/a}^\infty ae^{(u-b)s/a}F(u)du$$

where we have made the substitution $t = au + b$ so that $G(t) = F(u)$. In terms of $\bar{f}(as)$ this is

$$ae^{-sb}\bar{f}(as) + ae^{-sb}\int_{-b/a}^0 e^{-ast}F(t)dt.$$

8. The proof proceeds by using the definition as follows:

$$\mathcal{L}\{F(at)\} = \int_0^\infty e^{-st}F(at)dt = \int_0^\infty e^{-su/a}F(u)du/a$$

which gives the result. Evaluation of the two Laplace Transforms follows from using the results of Exercise 5 alongside the change of scale result just derived with, for (a) $a = 6$ and for (b) $a = 7$. The answers are

$$\text{(a)}\frac{36 - s^2}{(36 + s^2)^2}, \quad \text{(b)}\frac{14s^3 - 6.7^3s}{(49 + s^2)^3}.$$

Exercises 2.8

1. If $F(t) = \cos(at)$ then $F'(t) = -a\sin(at)$. The derivative formula thus gives

$$\mathcal{L}\{-a\sin(at)\} = s\mathcal{L}\{\cos(at)\} - F(0).$$

Assuming we know that $\mathcal{L}\{\cos(at)\} = \dfrac{s}{s^2 + a^2}$ then, straightforwardly

$$\mathcal{L}\{-a\sin(at)\} = s\frac{s}{s^2 + a^2} - 1 = -\frac{a^2}{s^2 + a^2}$$

i.e $\mathcal{L}\{\sin(at)\} = \dfrac{a}{s^2 + a^2}$ as expected.

2. Using Theorem 2.2 gives

$$\mathcal{L}\left\{\frac{\sin(t)}{t}\right\} = s\mathcal{L}\left\{\int_0^t \frac{\sin(u)}{u} du\right\}$$

In the text (after Theorem 2.4) we have derived that

$$\mathcal{L}\left\{\int_0^t \frac{\sin(u)}{u} du\right\} = \frac{1}{s}\tan^{-1}\left\{\frac{1}{s}\right\},$$

in fact this calculation is that one in reverse. The result

$$\mathcal{L}\left\{\frac{\sin(t)}{t}\right\} = \tan^{-1}\left\{\frac{1}{s}\right\}$$

is immediate. In order to derive the required result, the following manipulations need to take place:

$$\mathcal{L}\left\{\frac{\sin(t)}{t}\right\} = \int_0^\infty e^{-st}\frac{\sin(t)}{t} dt$$

and if we substitute $ua = t$ the integral becomes

$$\int_0^\infty e^{-asu}\frac{\sin(au)}{u} du.$$

This is still equal to $\tan^{-1}\left\{\dfrac{1}{s}\right\}$. Writing $p = as$ then gives the result. (p is a dummy variable of course that can be re-labelled s.)

3. The calculation is as follows:

$$\mathcal{L}\left\{\int_0^t p(v)dv\right\} = \frac{1}{s}\mathcal{L}\{p(v)\}$$

so

$$\mathcal{L}\left\{\int_0^t \int_0^v F(u)dudv\right\} = \frac{1}{s}\mathcal{L}\left\{\int_0^v F(u)du\right\} = \frac{1}{s^2}f(s)$$

as required.

4. Using Theorem 2.4 we get

$$\mathcal{L}\left\{\int_0^t \frac{\cos(au) - \cos(bu)}{u} du\right\} = \frac{1}{s}\int_s^\infty \frac{u}{a^2 + u^2} - \frac{u}{b^2 + u^2} du.$$

These integrals are standard "ln" and the result $\frac{1}{s}\ln\left(\frac{s^2 + a^2}{s^2 + b^2}\right)$ follows at once.

5. This transform is computed directly as follows

$$\mathcal{L}\left\{\frac{2\sin(t)\sinh(t)}{t}\right\} = \mathcal{L}\left\{\frac{e^t\sin(t)}{t}\right\} - \mathcal{L}\left\{\frac{e^{-t}\sin(t)}{t}\right\}.$$

Using the first shift theorem (Theorem 1.3) and the result of Exercise 2 above yields the result that the required Laplace Transform is equal to

$$\tan^{-1}\left(\frac{1}{s-1}\right) - \tan^{-1}\left(\frac{1}{s+1}\right) = \tan^{-1}\left(\frac{2}{s^2}\right).$$

(The identity $\tan^{-1}(x) - \tan^{-1}(y) = \tan^{-1}\left(\frac{x-y}{1+xy}\right)$ has been used.)

6. This follows straight from the definition of Laplace Transform:

$$\lim_{s\to\infty} \bar{f}(s) = \lim_{s\to\infty}\int_0^\infty e^{-st}F(t)dt = \int_0^\infty \lim_{s\to\infty} e^{-st}F(t)dt = 0.$$

It also follows from the final value theorem (Theorem 2.13) in that if $\lim_{s\to\infty} s\bar{f}(s)$ is finite then by necessity $\lim_{s\to\infty} \bar{f}(s) = 0$.

7. These problems are all reasonably straightforward

(a) $$\frac{2(2s+7)}{(s+4)(s+2)} = \frac{3}{s+2} + \frac{1}{s+4}$$

and inverting each Laplace Transform term by term gives the result $3e^{-2t} + e^{-4t}$

(b) Similarly $$\frac{s+9}{s^2-9} = \frac{2}{s-3} - \frac{1}{s+3}$$

and the result of inverting each term gives $2e^{3t} - e^{-3t}$

(c) $$\frac{s^2 + 2k^2}{s(s^2 + 4k^2)} = \frac{1}{2}\left(\frac{1}{s} + \frac{s}{s^2 + 4k^2}\right)$$

and inverting gives the result

$$\frac{1}{2} + \frac{1}{2}\cos(2kt) = \cos^2(kt).$$

$$\text{(d)} \quad \frac{1}{s(s+3)^2} = \frac{1}{9s} - \frac{1}{9(s+3)} - \frac{1}{3(s+3)^2}$$

which inverts to

$$\frac{1}{9} - \frac{1}{9}(3t+1)e^{-3t}.$$

(d) This last part is longer than the others. The partial fraction decomposition is best done by computer algebra, although hand computation is possible. The result is

$$\frac{1}{(s-2)^2(s+3)^3} = \frac{1}{125(s-2)^2} - \frac{3}{625(s-2)} + \frac{1}{25(s+3)^3} + \frac{2}{125(s+3)^2}$$
$$+ \frac{3}{625(s+3)}$$

and the inversion gives $\dfrac{e^{2t}}{625}(5t-3) + \dfrac{e^{-3t}}{1250}(25t^2 + 20t + 6).$

8. (a) $F(t) = 2 + \cos(t) \to 3$ as $t \to 0$, and as $\dfrac{2}{s} + \dfrac{s}{s^2+1}$ we also have that $sf(s) \to 2 + 1 = 3$ as $s \to \infty$ hence verifying the initial value theorem.
(b) $F(t) = (4+t)^2 \to 16$ as $t \to 0$. In order to find the Laplace Transform, we expand and evaluate term by term so that $sf(s) = 16 + 8/s + 2/s^2$ which obviously also tends to 16 as $s \to \infty$ hence verifying the theorem once more.

9. (a) $F(t) = 3 + e^{-t} \to 3$ as $t \to \infty$. $f(s) = \dfrac{3}{s} + \dfrac{1}{s+1}$ so that $sf(s) \to 3$ as $s \to 0$ as required by the final value theorem.
(b) With $F(t) = t^3 e^{-t}$, we have $f(s) = 6/(s+1)^4$ and as $F(t) \to 0$ as $t \to \infty$ and $sf(s)$ also tends to the limit 0 as $s \to 0$ the final value theorem is verified.

10. For small enough t, we have that

$$\sin(\sqrt{t}) = \sqrt{t} + O(t^{3/2})$$

and using the standard form (or the result on p50):

$$\mathcal{L}\{t^{x-1}\} = \frac{\Gamma\{x\}}{s^x}$$

with $x = 3/2$ gives

$$\mathcal{L}\{\sin(\sqrt{t})\} = \mathcal{L}\{\sqrt{t}\} + \cdots = \frac{\Gamma\{3/2\}}{s^{3/2}} + \cdots$$

and using that $\Gamma\{3/2\} = (1/2)\Gamma\{1/2\} = \sqrt{\pi}/2$ we deduce that

$$\mathcal{L}\{\sin(\sqrt{t})\} = \frac{\sqrt{\pi}}{2s^{3/2}} + \cdots.$$

Also, using the formula given,

$$\frac{k}{s^{3/2}} \exp{-\frac{1}{4s}} = \frac{k}{s^{3/2}} + \cdots.$$

Comparing these series for large values of s, equating coefficients of $s^{-3/2}$ gives

$$k = \frac{\sqrt{\pi}}{2}.$$

11. Using the power series expansions for sin and cos gives

$$\sin(t^2) = \sum_{n=0}^{\infty}(-1)^n \frac{t^{4n+2}}{(2n+1)!}$$

and

$$\cos(t^2) = \sum_{n=0}^{\infty}(-1)^n \frac{t^{4n}}{2n!}.$$

Taking the Laplace Transform term by term gives

$$\mathcal{L}\{\sin(t^2)\} = \sum_{n=0}^{\infty}(-1)^n \frac{(4n+2)!}{(2n+1)!s^{4n+3}}$$

and

$$\mathcal{L}\{\cos(t^2)\} = \sum_{n=0}^{\infty}(-1)^n \frac{(4n)!}{(2n)!s^{4n+1}}.$$

12. Given that $Q(s)$ is a polynomial with n distinct zeros, we may write

$$\frac{P(s)}{Q(s)} = \frac{A_1}{s-a_1} + \frac{A_2}{s-a_2} + \cdots + \frac{A_k}{s-a_k} + \cdots + \frac{A_n}{s-a_n}$$

where the A_ks are some real constants to be determined. Multiplying both sides by $s - a_k$ then letting $s \to a_k$ gives

$$A_k = \lim_{s \to a_k} \frac{P(s)}{Q(s)}(s - a_k) = \lim_{s \to a_k} P(s)\frac{(s-a_k)}{Q(s)}.$$

Using l'Hôpital's rule now gives

$$A_k = \frac{P(a_k)}{Q'(a_k)}$$

for all $k = 1, 2, \ldots, n$. This is true for all k, thus we have established that

$$\frac{P(s)}{Q(s)} = \frac{P(a_1)}{Q'(a_1)}\frac{1}{(s-a_1)} + \cdots + \frac{P(a_k)}{Q'(a_k)}\frac{1}{(s-a_k)} + \cdots \frac{P(a_n)}{Q'(a_n)}\frac{1}{(s-a_n)}.$$

Taking the inverse Laplace Transform gives the result

$$\mathcal{L}^{-1}\left\{\frac{P(s)}{Q(s)}\right\} = \sum_{k=1}^{n}\frac{P(a_k)}{Q'(a_k)}e^{a_k t}$$

sometimes known as Heaviside's expansion formula.

13. All of the problems in this question are solved by evaluating the Laplace Transform explicitly.

(a) $\mathcal{L}\{H(t-a)\} = \int_a^\infty e^{-st}dt = \dfrac{e^{-as}}{s}.$

(b) $\mathcal{L}\{f_1(t)\} = \int_0^2 (t+1)e^{-st}dt + \int_2^\infty 3e^{-st}dt.$

Evaluating the right-hand integrals gives the solution

$$\frac{1}{s} + \frac{1}{s^2}(e^{-2s} - 1).$$

(c) $\mathcal{L}\{f_2(t)\} = \int_0^2 (t+1)e^{-st}dt + \int_2^\infty 6e^{-st}dt.$

Once again, evaluating gives

$$\frac{1}{s} + \frac{3}{s}e^{-2s} + \frac{1}{s^2}(e^{-2s} - 1)$$

(d) As the function $f_1(t)$ is in fact continuous in the interval $[0, \infty)$ the formula for the derivative of the Laplace Transform (Theorem 2.2) can be used to give the result $\dfrac{1}{s}(e^{-2s} - 1)$ at once. Alternative, f_1 can be differentiated (it is $1 - H(t - 2)$) and evaluated directly.

14. We use the formula for the Laplace Transform of a periodic function Theorem 2.19 to give

$$\mathcal{L}\{F(t)\} = \frac{\int_0^{2c} e^{-st}F(t)dt}{(1 - e^{2sc})}.$$

The numerator is evaluated directly:

$$\int_0^{2c} e^{-st}F(t)dt = \int_0^c te^{-st}dt + \int_c^{2c} (2c - t)e^{-st}dt$$

which after routine integration by parts simplifies to

$$\frac{1}{s^2}(e^{-sc} - 1)^2.$$

The Laplace Transform is thus

$$\mathcal{L}\{F(t)\} = \frac{1}{1 - e^{2sc}}\frac{1}{s^2}(e^{-sc} - 1)^2 = \frac{1}{s^2}\frac{1 - e^{-sc}}{1 + e^{sc}}$$

which simplifies to

$$\frac{1}{s^2}\tanh\left(\frac{1}{2}sc\right).$$

Exercises 3.6

1. (a) If we substitute $u = t - \tau$ into the definition of convolution then

$$g * f = \int_0^t g(\tau)f(t - \tau)d\tau$$

becomes

$$-\int_t^0 g(u - \tau)f(u)du = g * f.$$

(b) Associativity is proved by effecting the transformation $(u, \tau) \rightarrow (x, y)$ where $u = t - x - y$, and $\tau = y$ on the expression

$$f * (g * h) = \int_0^t \int_0^{t-\tau} f(\tau)g(u)h(t - \tau - u)dud\tau.$$

The area covered by the double integral does not change under this transformation, it remains the right-angled triangle with vertices $(0, t)$, $(0, 0)$ and $(t, 0)$. The calculation proceeds as follows:

$$dud\tau = \frac{\partial(u, \tau)}{\partial(\tau, y)}dxdy = -dxdy$$

so that

$$f * (g * h) = \int_0^t \int_0^{t-x} f(y)g(t - x - y)h(x)dydx$$

$$= \int_0^t h(x) \left[\int_0^{t-x} f(y)g(t - x - y)dy \right] dx$$

$$= \int_0^t h(x)[f * g](t - x)dx = h * (f * g)$$

and this is $(f * g) * h$ by part (a) which establishes the result.
(c) Taking the Laplace Transform of the expression $f * f^{-1} = 1$ gives

$$\mathcal{L}\{f\}.\mathcal{L}\{f^{-1}\} = \frac{1}{s}$$

from which

$$\mathcal{L}\{f^{-1}\} = \frac{1}{s\bar{f}(s)}$$

using the usual notation ($\bar{f}(s)$ is the Laplace Transform of $f(t)$). It must be the case that $\dfrac{1}{s\bar{f}(s)} \rightarrow 0$ as $s \rightarrow \infty$. The function f^{-1} is not uniquely defined. Using the properties of the Dirac-δ function, we can also write

$$\int_0^{t+} f(\tau)\delta(t - \tau)d\tau = f(t)$$

from which

$$f^{-1}(t) = \frac{\delta(t-\tau)}{f(t)}.$$

Clearly, $f(t) \neq 0$.

2. Since $\mathcal{L}\{f\} = \bar{f}$ and $\mathcal{L}\{1\} = 1/s$ we have

$$\mathcal{L}\{f * 1\} = \frac{\bar{f}}{s}$$

so that, on inverting

$$\mathcal{L}^{-1}\left\{\frac{\bar{f}}{s}\right\} = f * 1 = \int_0^t f(\tau)d\tau$$

as required.

3. These convolution integrals are straightforward to evaluate:

$$\text{(a)} \quad t * \cos(t) = \int_0^t (t-\tau)\cos(\tau)d\tau$$

this is, using integration by parts

$$1 - \cos(t).$$

$$\text{(b)} \quad t * t = \int_0^t (t-\tau)\tau d\tau = \frac{t^3}{6}.$$

$$\text{(c)} \quad \sin(t) * \sin(t) = \int_0^t \sin(t-\tau)\sin(\tau)d\tau = \frac{1}{2}\int_0^t [\cos(2\tau - t) - \cos(t)]d\tau$$

this is now straightforwardly

$$\frac{1}{2}(\sin(t) + t\cos(t)).$$

$$\text{(d)} \quad e^t * t = \int_0^t e^{t-\tau}\tau d\tau$$

which on integration by parts gives

$$-1 - t + e^{-t}.$$

$$\text{(e)} \quad e^t * \cos(t) = \int_0^t e^{t-\tau}\cos(\tau)d\tau.$$

Integration by parts twice yields the following equation

$$\int_0^t e^{t-\tau}\cos(\tau)d\tau = \left[e^{-\tau}\sin(\tau) - e^{-\tau}\cos(\tau)\right]_0^t - \int_0^t e^{t-\tau}\cos(\tau)d\tau$$

from which

$$\int_0^t e^{t-\tau}\cos(\tau)d\tau = \frac{1}{2}(\sin(t) - \cos(t) + e^t).$$

4. (a) This is proved by using l'Hôpital's rule as follows

$$\lim_{x \to 0} \left\{ \frac{\mathrm{erf}(x)}{x} \right\} = \lim_{x \to 0} \frac{1}{x} \frac{2}{\sqrt{\pi}} \int_0^x e^{-t^2} dt = \frac{2}{\sqrt{\pi}} \frac{d}{dx} \int_0^x e^{-t^2} dt$$

and using Leibnitz' rule (or differentiation under the integral sign) this is

$$\lim_{x \to 0} \frac{2}{\sqrt{\pi}} e^{-x^2} = \frac{2}{\sqrt{\pi}}$$

as required.

(b) This part is tackled using power series expansions. First note that

$$e^{-x^2} = 1 - x^2 + \frac{x^4}{2!} - \frac{x^6}{3!} + \cdots + (-1)^{n+1} \frac{x^{2n}}{n!} + \cdots.$$

Integrating term by term (uniformly convergent for all x) gives

$$\int_0^{\sqrt{t}} e^{-x^2} dx = t^{1/2} - \frac{t^{3/2}}{3} + \frac{t^{5/2}}{5.2!} - \frac{t^{7/2}}{7.3!} + \cdots + (-1)^{n+1} \frac{t^{n+1/2}}{(2n+1).n!} + \cdots$$

from which

$$t^{-\frac{1}{2}} \mathrm{erf}(\sqrt{t}) = \frac{2}{\sqrt{\pi}} \left(1 - \frac{t}{3} + \frac{t^2}{5.2!} - \frac{t^3}{7.3!} + \cdots + (-1)^{n+1} \frac{t^n}{(2n+1).n!} + \cdots \right).$$

Taking the Laplace Transform of this series term by term (again justified by the uniform convergence of the series for all t) gives

$$\mathcal{L}^{-1}\{t^{-\frac{1}{2}} \mathrm{erf}(\sqrt{t})\} = \frac{2}{\sqrt{\pi}} \left(\frac{1}{s} - \frac{1}{3s^2} + \frac{1}{5s^3} - \frac{1}{7s^4} + \cdots + \frac{(-1)^n}{(2n+1)s^{n+1}} + \cdots \right)$$

and taking out a factor $1/\sqrt{s}$ leaves the arctan series for $1/\sqrt{s}$. Hence we get the required result:

$$\mathcal{L}^{-1}\{t^{-1/2} \mathrm{erf}(\sqrt{t})\} = \frac{2}{\sqrt{\pi s}} \tan^{-1}\left(\frac{1}{\sqrt{s}} \right).$$

5. All of these differential equations are solved by taking Laplace Transforms. Only some of the more important steps are shown.

(a) The transformed equation is

$$s\bar{x}(s) - x(0) + 3\bar{x}(s) = \frac{1}{s-2}$$

from which, after partial fractions,

$$\bar{x}(s) = \frac{1}{s+3} + \frac{1}{(s-2)(s+3)} = \frac{4/5}{s+3} + \frac{1/5}{s-2}.$$

Inverting gives

$$x(t) = \frac{4}{5}e^{-3t} + \frac{1}{5}e^{2t}.$$

(b) This equation has Laplace Transform

$$(s+3)\bar{x}(s) - x(0) = \frac{1}{s^2+1}$$

from which

$$\bar{x}(s) = \frac{x(0)}{s+3} - \frac{1/10}{s+3} + \frac{s/10 - 3/10}{s^2+1}.$$

The boundary condition $x(\pi) = 1$ is not natural for Laplace Transforms, however inverting the above gives

$$x(t) = \left(x(0) - \frac{1}{10}\right)e^{-3t} - \frac{1}{10}\cos(t) + \frac{3}{10}\sin(t)$$

and this is 1 when $x = \pi$, from which

$$x(0) - \frac{1}{10} = \frac{9}{10}e^{3\pi}$$

and the solution is

$$x(t) = \frac{9}{10}e^{3(\pi-t)} - \frac{1}{10}\cos(t) + \frac{3}{10}\sin(t).$$

(c) This equation is second order; the principle is the same but the algebra is messier. The Laplace Transform of the equation is

$$s^2\bar{x}(s) + 4s\bar{x}(s) + 5\bar{x}(s) = \frac{8}{s^2+1}$$

and rearranging using partial fractions gives

$$\bar{x}(s) = \frac{s+2}{(s+2)^2+1} + \frac{1}{(s+2)^2+1} - \frac{s}{s^2+1} + \frac{1}{s^2+1}.$$

Taking the inverse then yields the result

$$x(t) = e^{-2t}(\cos(t) + \sin(t)) + \sin(t) - \cos(t).$$

(d) The Laplace Transform of the equation is

$$(s^2 - 3s - 2)\bar{x}(s) - s - 1 + 3 = \frac{6}{s}$$

from which, after rearranging and using partial fractions,

$$\bar{x}(s) = -\frac{3}{s} + \frac{4(s - \frac{3}{2})}{(s - \frac{3}{2})^2 - \frac{17}{4}} - \frac{5}{(s - \frac{3}{2})^2 - \frac{17}{4}}$$

which gives the solution

$$x(t) = -3 + 4e^{\frac{3}{2}t}\cosh\left(\frac{t}{2}\sqrt{17}\right) - \frac{10}{\sqrt{17}}e^{\frac{3}{2}t}\sinh\left(\frac{t}{2}\sqrt{17}\right).$$

(e) This equation is solved in a similar way. The transformed equation is

$$s^2\bar{y}(s) - 3s + \bar{y}(s) - 1 = \frac{6}{s^2 + 4}$$

from which

$$\bar{y}(s) = -\frac{2}{s^2 + 4} + \frac{3s + 3}{s^2 + 1}$$

and inverting, the solution

$$y(t) = -\sin(2t) + 3\cos(t) + 3\sin(t)$$

results.

6. Simultaneous ODEs are transformed into simultaneous algebraic equations and the algebra to solve them is often horrid. For parts (a) and (c) the algebra can be done by hand, for part (b) computer algebra is almost compulsory!

(a) The simultaneous equations in the transformed state after applying the boundary conditions are

$$(s - 2)\bar{x}(s) - (s + 1)\bar{y}(s) = \frac{6}{s - 3} + 3$$

$$(2s - 3)\bar{x}(s) + (s - 3)\bar{y}(s) = \frac{6}{s - 3} + 6$$

from which we solve and rearrange to obtain

$$\bar{x}(s) = \frac{4}{(s - 3)(s - 1)} + \frac{3s - 1}{(s - 1)^2}$$

so that, using partial fractions

$$\bar{x}(s) = \frac{2}{s - 3} + \frac{1}{s - 1} + \frac{2}{(s - 1)^2}$$

giving, on inversion

$$x(t) = 2e^{3t} + e^t + 2te^t.$$

In order to find $y(t)$ we eliminate dy/dt from the original pair of simultaneous ODEs to give

$$y(t) = -3e^{3t} - \frac{5}{4}x(t) + \frac{3}{4}\frac{dx}{dt}.$$

Substituting for $x(t)$ then gives

$$y(t) = -e^{3t} + e^t - te^t.$$

(b) This equation is most easily tackled by substituting the derivative of

$$y = -4\frac{dx}{dt} - 6x + 2\sin(2t)$$

into the second equation to give

$$5\frac{d^2}{dx^2} + 6\frac{dx}{dt} + x = 4\cos(2t) + 3e^{-2t}.$$

The Laplace Transform of this is then

$$5(s^2\bar{x}(s) - sx(0) - x'(0)) + 6(s\bar{x}(s) - x(0)) + \bar{x}(s) = \frac{4s}{s^2+4} + \frac{3}{s+2}.$$

After inserting the given boundary conditions and rearranging we are thus face with inverting

$$\bar{x}(s) = \frac{10s+2}{5s^2+6s+1} + \frac{4s}{(s^2+4)(5s^2+6s+1)} + \frac{3}{(s+2)(5s^2+6s+1)}.$$

Using a partial fractions package gives

$$\bar{x}(s) = \frac{29}{20(s+1)} + \frac{1}{3(s+2)} + \frac{2225}{1212(5s+1)} - \frac{4(19s-24)}{505(s^2+4)}$$

and inverting yields

$$x(t) = \frac{1}{3}e^{-2t} + \frac{29}{20}e^{-t} + \frac{445}{1212}e^{-\frac{1}{5}t} - \frac{76}{505}\cos(2t) + \frac{48}{505}\sin(2t).$$

Substituting back for $y(t)$ gives

$$y(t) = \frac{2}{3}e^{2t} - \frac{29}{10}e^{-t} - \frac{1157}{606}e^{-\frac{1}{5}t} + \frac{72}{505}\cos(2t) + \frac{118}{505}\sin(2t).$$

(c) This last problem is fully fourth order, but we do not change the line of approach. The Laplace transform gives the simultaneous equations

$$(s^2 - 1)\bar{x}(s) + 5s\bar{y}(s) - 5 = \frac{1}{s^2}$$

$$-2s\bar{x}(s) + (s^2 - 4)\bar{y}(s) - s = -\frac{2}{s}$$

in which the boundary conditions have already been applied. Solving for $\bar{y}(s)$ gives

$$\bar{y}(s) = \frac{s^4 + 7s^2 + 4}{s(s^2+4)(s^2+1)} = \frac{1}{s} - \frac{2}{3}\frac{s}{s^2+4} + \frac{2}{3}\frac{s}{s^2+1}$$

which inverts to the solution

$$y(t) = 1 - \frac{2}{3}\cos(2t) + \frac{2}{3}\cos(t).$$

Substituting back into the second original equation gives

$$x(t) = -t - \frac{5}{3}\sin(t) + \frac{4}{3}\sin(2t).$$

7. Using Laplace Transforms, the transform of x is given by

$$\bar{x}(s) = \frac{A}{(s^2+1)(s^2+k^2)} + \frac{v_0}{(s^2+k^2)} + \frac{sx(0)}{(s^2+k^2)}.$$

If $k \neq 1$ this inverts to

$$x(t) = \frac{A}{k^2-1}\left(\sin(t) - \frac{\sin(kt)}{k}\right) + \frac{v_0}{k}\sin(kt) + x_0\cos(kt).$$

If $k = 1$ there is a term $(1+s^2)^2$ in the denominator, and the inversion can be done using convolution. The result is

$$x(t) = \frac{A}{2}(\sin(t) - t\cos(t)) + v_0\sin(t) + x(0)\cos(t)$$

and it can be seen that this tends to infinity as $t \to \infty$ due to the term $t\cos(t)$. This is called a *secular* term. It is not present in the solution for $k \neq 1$ which is purely oscillatory. The presence of a secular term denotes resonance.

8. Taking the Laplace Transform of the equation, using the boundary conditions and rearranging gives

$$\bar{x}(s) = \frac{sv_0 + g}{s^2(s+a)}$$

which after partial fractions becomes

$$\bar{x}(s) = \frac{-\frac{1}{a^2}(av_0 - g)}{s+a} + \frac{-\frac{1}{a^2}(av_0 - g)s + \frac{g}{a}}{s^2}.$$

This inverts to the expression in the question. The speed

$$\frac{dx}{dt} = \frac{g}{a} - \frac{(av_0 - g)}{a}e^{-at}.$$

As $t \to \infty$ this tends to g/a which is the required terminal speed.

9. The set of equations in matrix form is determined by taking the Laplace Transform of each. The resulting algebraic set is expressed in matrix form as follows:

$$\begin{pmatrix} 1 & -1 & -1 & 0 \\ R_1 & sL_2 & 0 & 0 \\ R_1 & 0 & R_3 & 1/C \\ 0 & 0 & 1 & -s \end{pmatrix} \begin{pmatrix} \bar{j}_1 \\ \bar{j}_2 \\ \bar{j}_3 \\ \bar{q}_3 \end{pmatrix} = \begin{pmatrix} 0 \\ L_2 j_2(0) + E\omega/(\omega^2 + s^2) \\ E\omega/(\omega^2 + s^2) \\ -q_3(0) \end{pmatrix}.$$

10. The Laplace Transform of this fourth order equation is

$$k(s^4\bar{y}(s) - s^3 y(0) - s^2 y'(0) - sy''(0) - y'''(0)) = \frac{\omega_0}{c}\left(\frac{c}{s} - \frac{1}{s^2} + \frac{e^{-as}}{s^2}\right)$$

Using the boundary conditions is easy for those given at $x = 0$, the others give

$$y''(0) - 2cy'''(0) + \frac{5}{6}\omega_0 c^2 = 0 \quad \text{and} \quad y'''(0) = \frac{1}{2}\omega_0 c.$$

So $y''(0) = \frac{1}{6}\omega_0 c^2$ and the full solution is, on inversion

$$y(x) = \frac{1}{12}\omega_0 c^2 x^2 - \frac{1}{12}\omega_0 c x^3 + \frac{\omega_0}{120}\left[5cx^4 - x^5 + (x - c)^5 H(x - c)\right]$$

where $0 \leq x \leq 2c$. Differentiating twice and putting $x = c/2$ gives
$$y''(c/2) = \frac{1}{48}\omega_0 c^2.$$

11. Taking the Laplace Transform and using the convolution theorem gives

$$\bar{\phi}(s) = \frac{2}{s^3} + \bar{\phi}(s)\frac{1}{s^2 + 1}$$

from which

$$\bar{\phi}(s) = \frac{2}{s^5} + \frac{2}{s^2}.$$

Inversion gives the solution

$$\phi(t) = t^2 + \frac{1}{12}t^4.$$

Exercises 4.7

1. The Riemann–Lebesgue lemma is stated as Theorem 4.2. As the constants b_n in a Fourier sine series for $g(t)$ in $[0, \pi]$ are given by

$$b_n = \frac{2}{\pi}\int_0^\pi g(t)\sin(nt)dt$$

and these sine functions form a basis for the linear space of piecewise continuous functions in $[0, \pi]$ (with the usual inner product) of which $g(t)$ is a member, the Riemann–Lebesgue lemma thus immediately gives the result. More directly, Parseval's Theorem:

$$\int_{-\pi}^{\pi}[g(t)]^2 dt = \pi a_0^2 + \pi \sum_{n=1}^{\infty}(a_n^2 + b_n^2)$$

yields the results

$$\lim_{n\to\infty}\int_{-\pi}^{\pi} g(t)\cos(nt)dt = 0$$

$$\lim_{n\to\infty}\int_{-\pi}^{\pi} g(t)\sin(nt)dt = 0$$

as the nth term of the series on the right has to tend to zero as $n \to \infty$. As $g(t)$ is piecewise continuous over the half range $[0, \pi]$ and is free to be defined as odd over the full range $[-\pi, \pi]$, the result follows.

2. The differential equation can be written

$$\frac{d}{dt}\left[(1-t^2)\frac{dP_n}{dt}\right] = -n(n+1)P_n.$$

This means that the integral can be manipulated using integration by parts as follows:

$$\int_{-1}^{1} P_m P_n dt = -\frac{1}{n(n+1)} \int_{-1}^{1} \frac{d}{dt}\left[(1-t^2)\frac{dP_n}{dt}\right] P_m dt$$

$$= -\frac{1}{n(n+1)} \left[(1-t^2)\frac{dP_n}{dt}\frac{dP_m}{dt}\right]_{-1}^{1}$$

$$+ \frac{1}{n(n+1)} \int_{-1}^{1} \frac{d}{dt}\left[(1-t^2)\frac{dP_m}{dt}\right] P_n dt$$

$$= \frac{1}{n(n+1)} \left[(1-t^2)\frac{dP_m}{dt}\frac{dP_n}{dt}\right]_{-1}^{1} + \frac{m(m+1)}{n(n+1)} \int_{-1}^{1} P_m P_n dt$$

$$= \frac{m(m+1)}{n(n+1)} \int_{-1}^{1} P_m P_n dt,$$

all the integrated bits being zero. Therefore

$$\int_{-1}^{1} P_m P_n dt = 0$$

unless $m = n$ as required.

3. The Fourier series is found using the formulae in Section 4.2. The calculation is routine if lengthy and the answer is

$$f(t) = \frac{5\pi}{16} - \frac{2}{\pi}\left(\cos(t) + \frac{\cos(3t)}{3^2} + \frac{\cos(5t)}{5^2} + \cdots\right)$$

$$- \frac{2}{\pi}\left(\frac{\cos(2t)}{2^2} + \frac{\cos(6t)}{6^2} + \frac{\cos(10t)}{10^2}\cdots\right)$$

$$+ \frac{1}{\pi}\left(\sin(t) - \frac{\sin(3t)}{3^2} + \frac{\sin(5t)}{5^2} - \frac{\sin(7t)}{7^2}\cdots\right).$$

This function is displayed in Figure A.1.

4. The Fourier series for $H(x)$ is found straightforwardly as

$$\frac{1}{2} + \frac{2}{\pi} \sum_{n=1}^{\infty} \frac{\sin(2n-1)x}{2n-1}.$$

Put $x = \pi/2$ and we get the series in the question and its sum:

$$1 - \frac{1}{3} + \frac{1}{5} - \frac{1}{7} + \cdots = \frac{\pi}{4}$$

a series attributed to the Scottish mathematician James Gregory (1638–1675).

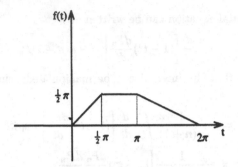

Figure A.1: The original function composed of straight line segments

5. The Fourier series has the value

$$f(x) \sim -\frac{8}{\pi} \sum_{n=1}^{\infty} \frac{n \sin(2nx)}{(2n+1)(2n-1)}.$$

6. This is another example where the Fourier series is found straightforwardly using integration by parts. The result is

$$1 - x^2 \sim \frac{1}{2}\left(\pi - \frac{\pi^3}{3}\right) - 4\sum_{n=1}^{\infty} \frac{(-1)^n}{n^2}\cos(nx).$$

As the Fourier series is in fact continuous for this example there is no controversy, at $x = \pi$, $f(x) = 1 - \pi^2$.

7. Evaluating the integrals takes a little stamina this time.

$$b_n = \frac{1}{\pi}\int_{-\pi}^{\pi} e^{ax}\sin(nx)dx$$

and integrating twice by parts gives

$$b_n = \left[\frac{1}{a\pi}\sin(nx) - \frac{n}{a^2\pi}\cos(nx)\right]_{-\pi}^{\pi} - \frac{n^2}{a^2}b_n$$

from which

$$b_n = -\frac{2n\sinh\left(a\pi(-1)^n\right)}{a^2\pi}, \quad n = 1, 2, \ldots.$$

Similarly,

$$a_n = \frac{2a\sinh\left(a\pi(-1)^n\right)}{a^2\pi}, \quad n = 1, 2, \ldots,$$

and

$$a_0 = \frac{2\sinh(\pi a)}{\pi a}.$$

This gives the series in the question. Putting $x = 0$ gives the equation

$$e^0 = 1 = \frac{\sinh(\pi a)}{\pi}\left\{\frac{1}{a} + 2\sum_{n=1}^{\infty}\frac{(-1)^n a}{n^2 + a^2}\right\}$$

from which

$$\sum_{n=1}^{\infty}\frac{(-1)^n}{n^2 + a^2} = \frac{1}{2a^2}(a\pi\text{cosech}\,(a\pi) - 1).$$

Also, since

$$\sum_{-\infty}^{\infty}\frac{(-1)^n}{n^2 + a^2} = 2\sum_{n=1}^{\infty}\frac{(-1)^n}{n^2 + a^2} + \frac{1}{a^2},$$

we get the result

$$\sum_{-\infty}^{\infty}\frac{(-1)^n}{n^2 + a^2} = \frac{\pi}{a}\text{cosech}\,(a\pi).$$

Putting $x = \pi$ and using Dirichlet's theorem (Theorem 4.5) we get

$$\frac{1}{2}(f(\pi) + f(-\pi)) = \cosh(a\pi) = \frac{\sinh(\pi a)}{\pi}\left\{\frac{1}{a} + 2\sum_{n=1}^{\infty}\frac{a}{n^2 + a^2}\right\}$$

from which

$$\sum_{n=1}^{\infty}\frac{1}{n^2 + a^2} = \frac{1}{2a^2}(a\pi\coth(a\pi) - 1).$$

Also, since

$$\sum_{-\infty}^{\infty}\frac{1}{n^2 + a^2} = 2\sum_{n=1}^{\infty}\frac{1}{n^2 + a^2} + \frac{1}{a^2}$$

we get the result

$$\sum_{-\infty}^{\infty}\frac{1}{n^2 + a^2} = \frac{\pi}{a}\coth(a\pi).$$

8. The graph is shown in Figure A.2 and the Fourier series itself is given by

$$f(t) = \frac{1}{2}\pi + \frac{1}{\pi}\sinh(\pi)$$

$$+ \frac{2}{\pi}\sum_{n=1}^{\infty}\left[\frac{(-1)^n - 1}{n^2} + \frac{(-1)^n \sinh(\pi)}{n^2 + 1}\right]\cos(nt)$$

$$- \frac{2}{\pi}\sum_{n=1}^{\infty}\frac{(-1)^n}{n^2 + 1}\sinh(\pi)\sin(nt).$$

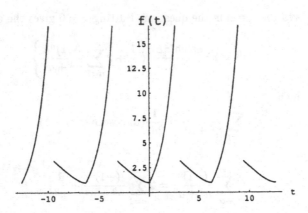

Figure A.2: The function $f(t)$

9. The Fourier series expansion over the range $[-\pi, \pi]$ is found by integration to be

$$f(t) = \frac{2}{3}\pi^2 + \sum_{n=1}^{\infty}\left[\frac{2}{n^2}\cos(nt) + \frac{(-1)^n}{n}\pi\sin(nt)\right] - \frac{4}{\pi}\sum_{n=1}^{\infty}\frac{\sin(2n-1)t}{(2n-1)^3}$$

and Figure A.3 gives a picture of it. The required series are found by first putting $t = 0$ which gives

$$\pi^2 = \frac{2}{3}\pi^2 + 2\sum_{n=1}^{\infty}\frac{1}{n^2}$$

from which

$$\sum_{n=1}^{\infty}\frac{1}{n^2} = \frac{\pi^2}{6}.$$

Putting $t = \pi$ gives, using Dirichlet's theorem (Theorem 4.5)

$$\frac{\pi^2}{2} = \frac{2}{3}\pi^2 - 2\sum_{n=1}^{\infty}\frac{(-1)^{n+1}}{n^2}$$

from which

$$\sum_{n=1}^{\infty}\frac{(-1)^{n+1}}{n^2} = \frac{\pi^2}{12}.$$

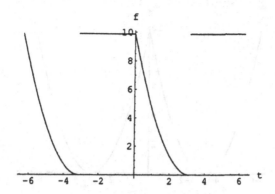

Figure A.3: The function $f(t)$ as a full Fourier series

10. The sine series is given by the formula

$$a_n = 0 \quad b_n = \frac{2}{\pi} \int_0^\pi (t - \pi)^2 \sin(nt)dt$$

with the result

$$f(t) \sim \frac{8}{\pi} \sum_{k=1}^\infty \frac{\sin(2k-1)t}{2k+1} + 2\pi \sum_{n=1}^\infty \frac{\sin(nt)}{n}.$$

This is shown in Figure A.5. The cosine series is given by

$$b_n = 0 \quad a_n = \frac{2}{\pi} \int_0^\pi (t - \pi)^2 \cos(nt)dt$$

from which

$$f(t) \sim -\frac{\pi^2}{3} + 4 \sum_{n=1}^\infty \frac{\cos(nt)}{n^2}$$

and this is pictured in Figure A.4.

11. Parseval's theorem (Theorem 4.19) is

$$\int_{-\pi}^\pi [f(t)]^2 dt = \pi a_0^2 + \pi \sum_{n=1}^\infty (a_n^2 + b_n^2).$$

Applying this to the Fourier series in the question is not straightforward, as we need the version for sine series. This is easily derived as

$$\int_0^\pi [f(t)]^2 dt = \frac{\pi}{2} \sum_{n=1}^\infty b_n^2.$$

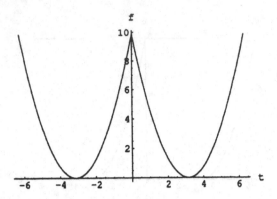

Figure A.4: The function $f(t)$ as an even function

The left hand side is

$$\int_0^\pi t^2(\pi - t)^2 dt = \frac{\pi^5}{30}.$$

The right hand side is

$$\frac{32}{\pi} \sum_{n=1}^\infty \frac{1}{(2n-1)^6}$$

whence the result

$$\sum_{n=1}^\infty \frac{1}{(2n-1)^6} = \frac{\pi^6}{960}.$$

Noting that

$$\sum_{n=1}^\infty \frac{1}{n^6} = \sum_{n=1}^\infty \frac{1}{(2n-1)^6} + \frac{1}{2^6} \sum_{n=1}^\infty \frac{1}{n^6}$$

gives the result

$$\sum_{n=1}^\infty \frac{1}{n^6} = \frac{64}{63} \frac{\pi^6}{960} = \frac{\pi^6}{945}.$$

12. The Fourier series for the function x^4 is found as usual by evaluating the integrals

$$a_n = \frac{1}{\pi} \int_{-\pi}^\pi x^4 \cos(nx) dx$$

and

$$b_n = \frac{1}{\pi} \int_{-\pi}^\pi x^4 \sin(nx) dx.$$

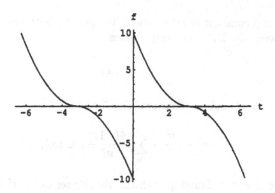

Figure A.5: The function $f(t)$ as an odd function

However as x^4 is an even function, $b_n = 0$ and there are no sine terms. Evaluating the integral for the cosine terms gives the series in the question. With $x = 0$, the series becomes

$$0 = \frac{\pi^4}{5} + 8\pi^2 \sum_{n=1}^{\infty} \frac{(-1)^n}{n^2} - 48 \sum_{n=1}^{\infty} \frac{(-1)^{n+1}}{n^4}$$

and using the value of the second series $(\pi^2)/12$ gives

$$\sum_{n=1}^{\infty} \frac{(-1)^{n+1}}{n^4} = \frac{7\pi^4}{15} \frac{1}{48} = \frac{7\pi^4}{720}.$$

Differentiating term by term yields the Fourier series for x^3 for $-\pi < x < \pi$ as

$$x^3 \sim \sum_{n=1}^{\infty} (-1)^n \frac{2}{n^3} (6 - \pi^2 n^2) \sin(nx).$$

13. Integrating the series

$$x \sim \sum_{n=1}^{\infty} \frac{2}{n} (-1)^{n+1} \sin(nx)$$

term by term gives

$$\frac{x^2}{2} \sim \sum_{n=1}^{\infty} \frac{2}{n^2} (-1)^n \cos(nx) + A$$

where A is a constant of integration. Integrating both sides with respect to x between the limits $-\pi$ and π gives

$$\frac{\pi^3}{3} = 2A\pi.$$

Hence $A = \pi^2/6$ and the Fourier series for x^2 is

$$x^2 \sim \frac{\pi^2}{3} + \sum_{n=1}^{\infty} \frac{4(-1)^n}{n^2} \cos(nx),$$

where $-\pi < x < \pi$. Starting with the Fourier series for x^4 over the same range and integrating term by term we get

$$\frac{x^5}{5} \sim \frac{\pi^4 x}{5} + \sum_{n=1}^{\infty} \frac{8(-1)^n}{n^5} (\pi^2 n^2 - 6) \sin(nx) + B$$

where B is the constant of integration. This time setting $x = 0$ immediately gives $B = 0$, but there is an x on the right-hand side that has to be expressed in terms of a Fourier series. We thus use

$$x \sim \sum_{n=1}^{\infty} \frac{2}{n} (-1)^{n+1} \sin(nx)$$

to give the Fourier series for x^5 in $[-\pi, \pi]$ as

$$x^5 \sim \sum_{n=1}^{\infty} (-1)^n \left[\frac{40\pi^2}{n^3} - \frac{240}{n^5} - \frac{2\pi^4}{n} \right] \sin(nx).$$

14. Using the complex form of the Fourier series, we have that

$$V(t) = \sum_{n=-\infty}^{\infty} c_n e^{2n\pi it/5}.$$

The coefficients are given by the formula

$$c_n = \frac{1}{5} \int_0^5 V(t) e^{2in\pi t/5} dt.$$

By direct calculation they are

$$c_{-4} = \frac{5}{\pi} \left(i + e^{7\pi i/10} \right)$$

$$c_{-3} = \frac{20}{3\pi} \left(i - e^{9\pi i/10} \right)$$

$$c_{-2} = \frac{10}{\pi} \left(i - e^{\pi i/10} \right)$$

$$c_{-1} = \frac{20}{\pi}\left(i + e^{3\pi i/10}\right)$$

$$c_0 = 16$$

$$c_1 = \frac{20}{\pi}\left(-i - e^{7\pi i/10}\right)$$

$$c_2 = \frac{10}{\pi}\left(-i + e^{9\pi i/10}\right)$$

$$c_3 = \frac{20}{3\pi}\left(-i + e^{\pi i/10}\right)$$

$$c_4 = \frac{10}{\pi}\left(-i - e^{3\pi i/10}\right).$$

Exercises 5.6

1. Using the separation of variable technique with

$$\phi(x,t) = \sum_k X_k(x)T_k(t)$$

gives the equations for $X_k(x)$ and $T_k(t)$ as

$$\frac{T_k'}{T_k} = \frac{\kappa X_k''}{X_k} = -\alpha^2$$

where $-\alpha^2$ is the separation constant. It is negative because $T_k(t)$ must not grow with t. In order to satisfy the boundary conditions we express $x(\pi/4 - x)$ as a Fourier sine series in the interval $0 \le x \le \pi/4$ as follows

$$x\left(x - \frac{\pi}{4}\right) = \frac{1}{2\pi}\sum_{k=1}^{\infty}\frac{1}{(2k-1)^3}\sin[4(2k-1)x].$$

Now the function $X_k(x)$ is identified as

$$X_k(x) = \frac{1}{2\pi}\frac{1}{(2k-1)^3}\sin[4(2k-1)x]$$

so that the equation obeyed by $X_k(x)$ together with the boundary conditions $\phi(0,t) = \phi(\pi/4,t) = 0$ for all time are satisfied. Thus

$$\alpha = \frac{4(2k-1)\pi}{\sqrt{\kappa}}.$$

Putting this expression for $X_k(x)$ together with

$$T_k(t) = e^{-16(2k-1)^2\pi^2 t/\kappa}$$

gives the solution

$$\phi(x,t) = \frac{1}{2\pi}\sum_{k=1}^{\infty}\frac{1}{(2k-1)^3}e^{-16(2k-1)^2\pi^2 t/\kappa}\sin[4(2k-1)x].$$

2. The equation is

$$a\frac{\partial^2 \phi}{\partial x^2} - b\frac{\partial \phi}{\partial x} - \frac{\partial \phi}{\partial t} = 0.$$

Taking the Laplace transform (in t) gives the ODE

$$a\bar{\phi}'' - b\bar{\phi}' - s\bar{\phi} = 0$$

after applying the boundary condition $\phi(x, 0) = 0$. Solving this and noting that

$$\phi(0, t) = 1 \Rightarrow \bar{\phi}(0, s) = \frac{1}{s} \quad \text{and} \quad \bar{\phi}(x, s) \to 0 \text{ as } x \to \infty$$

gives the solution

$$\bar{\phi} = \frac{1}{s} \exp \left\{ \frac{x}{2a} [b - \sqrt{b^2 + 4as}] \right\}.$$

3. Taking the Laplace transform of the heat conduction equation results in the ODE

$$\kappa \frac{d^2 \bar{\phi}}{dx^2} - s\bar{\phi} = -x \left(\frac{\pi}{4} - x \right).$$

Although this is a second order non-homogeneous ODE, it is amenable to standard complimentary function, particular integral techniques. The general solution is

$$\bar{\phi}(x, s) = A \cosh \left(x \sqrt{\frac{s}{\kappa}} \right) + B \sinh \left(x \sqrt{\frac{s}{\kappa}} \right) - \frac{x^2}{s} + \frac{\pi x}{4s} - \frac{2\kappa}{s^2}.$$

The inverse of this gives the expression in the question. If this inverse is evaluated term by term using the series in Appendix A, the answer of Exercise 1 is not regained immediately. However, if the factor $2\kappa t$ is expressed as a Fourier series, then the series solution is the same as that in Exercise 1.

4. Taking the Laplace transform in y gives the ODE

$$\frac{d^2 \bar{\phi}}{dx^2} = s\bar{\phi}.$$

Solving this, and applying the boundary condition $\phi(0, y) = 1$ which transforms to

$$\bar{\phi}(0, s) = \frac{1}{s}$$

gives the solution

$$\bar{\phi}(x, s) = \frac{1}{s} e^{-x\sqrt{s}}$$

which inverts to

$$\phi(x, y) = \text{erfc} \left\{ \frac{x}{2\sqrt{y}} \right\}.$$

5. Using the transformation suggested in the question gives the equation obeyed by $\phi(x,t)$ as

$$\frac{1}{c^2}\frac{\partial^2 \phi}{\partial t^2} = \frac{\partial^2 \phi}{\partial x^2}.$$

This is the standard wave equation. To solve this using Laplace transform techniques, we transform in the variable t to obtain the equation

$$\frac{d^2 \bar{\phi}}{dx^2} - \frac{s^2}{c^2}\bar{\phi} = -\frac{s}{c^2}\cos(mx).$$

The boundary conditions for $\phi(x,t)$ are

$$\phi(x,0) = \cos(mx) \quad \text{and} \quad \phi'(x,0) = -\frac{k}{2}\phi(x,0) = -\frac{k}{2}\cos(mx).$$

This last condition arises since

$$u'(x,t) = \frac{k}{2}e^{kt/2}\phi(x,t) + e^{kt/2}\phi'(x,t).$$

Applying these conditions gives, after some algebra

$$\bar{\phi}(x,s) = \left[\frac{1}{s} - \frac{s - \frac{k}{2}}{s^2 + m^2 c^2}\cos(mx)\right]e^{-\frac{sx}{c}} + \frac{s - \frac{k}{2}}{s^2 + m^2 c^2}\cos(mx).$$

Using the second shift theorem (Theorem 2.5) we invert this to obtain

$$u = \begin{cases} e^{kt/2}(1 - \cos(mct - mx)\cos(mx) + \frac{k}{2mc}\sin(mct - mx)\cos(mx) \\ \quad + \cos(mct)\cos(mx) - \frac{k}{2mc}\sin(mct)\cos(mx)) \qquad t > x/c \\ e^{kt/2}(\cos(mct)\cos(mx) - \frac{k}{2mc}\sin(mct)\cos(mx)) \qquad t < x/c. \end{cases}$$

6. Taking the Laplace transform of the one dimensional heat conduction equation gives

$$s\bar{u} = c^2 \bar{u}_{xx}$$

as $u(x,0) = 0$. Solving this with the given boundary condition gives

$$\bar{u}(x,s) = \bar{f}(s)e^{-x\sqrt{s}/c}.$$

Using the standard form

$$\mathcal{L}^{-1}\{e^{-a\sqrt{s}}\} = \frac{a}{2\sqrt{\pi t^3}}e^{-a^2/4t}$$

gives, using the convolution theorem

$$u = \frac{x}{2}\int_0^t f(\tau)\sqrt{\frac{k}{\pi(t-\tau)^3}}e^{-x^2/4k(t-\tau)}d\tau.$$

When $f(t) = \delta(t)$, $u = \frac{x}{2}\sqrt{\frac{k}{\pi t^3}}e^{-x^2/4kt}$.

7. Assuming that the heat conduction equation applies gives that

$$\frac{\partial T}{\partial t} = \kappa \frac{\partial^2 T}{\partial x^2}$$

so that when transformed $\bar{T}(x, s)$ obeys the equation

$$s\bar{T}(x, s) - T(x, 0) = \kappa \frac{d^2 \bar{T}}{dx^2}(x, s).$$

Now, $T(x, 0) = 0$ so this equation has solution

$$\bar{T}(x, s) = \bar{T}_0 e^{-x\sqrt{\frac{s}{\kappa}}}$$

and since

$$\frac{d\bar{T}}{dx}(s, 0) = -\frac{\alpha}{s}$$

$$\bar{T}_0 = \alpha \sqrt{\frac{\kappa}{s}}$$

and the solution is using standard forms (or Chapter 3)

$$T(x, t) = \alpha \sqrt{\frac{\kappa}{\pi t}} e^{-x^2/4\kappa t}.$$

This gives, at $x = 0$ the desired form for $T(0, t)$. Note that for non-zero x the solution is not singular at $t = 0$.

8. Taking the Laplace transform of the wave equation given yields

$$s^2 \bar{u} - \frac{\partial u}{\partial t}(x, 0) = c^2 \frac{d^2 \bar{u}}{dx^2}$$

so that substituting the series

$$\bar{u} = \sum_{n=0}^{\infty} \frac{a_n(x)}{s^{n+k+1}}, \quad k \text{ integer}$$

as in Example 5.5 gives $k = 1$ all the odd powers are zero and

$$\frac{a_0}{s^2} + \frac{a_2}{s^4} + \frac{a_6}{s^6} + \cdots - \cos(x) = c^2 \left(\frac{a_0''}{s^2} + \frac{a_2''}{s^4} + \frac{a_6''}{s^6} + \cdots \right)$$

so that

$$a_0 = \cos(x), \quad a_2 = c^2 a_0'' = -c^2 \cos(x) \quad a_4 = c^4 \cos(x) \text{ etc.}$$

Hence

$$\bar{u}(x, s) = \cos(x) \left(\frac{1}{s^2} - \frac{c^2}{s^4} + \frac{c^4}{s^6} - \frac{c^6}{s^8} + \cdots \right)$$

which inverts term by term to

$$u(x,t) = \cos(x)\left(t - \frac{c^2 t^3}{3!} + \frac{c^4 t^5}{5!} + \cdots\right)$$

which in this case converges to the closed form solution

$$u = \frac{1}{c}\cos(x)\sin(ct).$$

9. For this problem we proceed similarly. Laplace transforms are taken and the equation to solve this time is

$$s^2 \bar{u} - su(x,0) = c^2 \frac{d^2 \bar{u}}{dx^2}.$$

Once more substituting the series

$$\bar{u} = \sum_{n=0}^{\infty} \frac{a_n(x)}{s^{n+k+1}}, \quad k \text{ integer}$$

gives this time

$$a_0 + \frac{a_1}{s} + \frac{a_2}{s^2} + \cdots - \cos(x) - c^2\left(a_0'' + \frac{a_1''}{s} + \frac{a_2''}{s^2} \right)$$

so that

$$a_0 = \cos(x), \quad a_1 = 0 \quad a_2 = c^2 a_0'' = -c^2\cos(x) \quad a_3 = 0 \quad a_4 = c^4\cos(x) \text{ etc.}$$

giving

$$\bar{u}(x,s) = \sum_{n=0}^{\infty} \frac{c^{2n}(-1)^n \cos(x)}{s^{2n+1}}.$$

Inverting term by term gives the answer

$$u(x,t) = \cos(x)\sum_{n=0}^{\infty}(-1)^n \frac{c^{2n}t^{2n}}{2n!}$$

which in fact in this instance converges to the result

$$u = \cos(x)\cos(ct).$$

Exercises 6.6

1. With the function $f(t)$ as defined, simple integration reveals that

$$F(\omega) = \int_{-T}^{0} ke^{-i\omega t}\,dt + \int_{0}^{T} -ke^{-i\omega t}\,dt$$

$$= k \left[-\frac{e^{-i\omega t}}{i\omega} \right]_{-T}^{0} + k \left[\frac{e^{-i\omega t}}{i\omega} \right]_{0}^{T}$$

$$= \frac{2k}{i\omega} [\cos(\omega T) - 1]$$

$$= \frac{4ik}{\omega} \sin^2 \left(\frac{1}{2} \omega T \right).$$

2. With $f(t) = e^{-t^2}$ the Fourier transform is

$$F(\omega) = \int_{-\infty}^{\infty} e^{-t^2} e^{-i\omega t} dt = \int_{-\infty}^{\infty} e^{-(t-\frac{1}{2}i\omega)^2} e^{-\frac{1}{4}\omega^2} dt.$$

Now although there is a complex number ($\frac{1}{2} i\omega$) in the integrand, the change of variable $u = t - \frac{1}{2} i\omega$ can still be made. The limits are actually changed to $-\infty - \frac{1}{2} i\omega$ and $\infty - \frac{1}{2} i\omega$ but this does not change its value so we have that

$$\int_{-\infty}^{\infty} e^{(t-\frac{1}{2}i\omega)^2} dt = \sqrt{\pi}.$$

Hence

$$F(\omega) = \sqrt{\pi} e^{-\frac{1}{4}\omega^2}.$$

3. Consider the Fourier transform of the square wave, Example 6.2. The inverse yields:

$$\frac{A}{\pi} \int_{-\infty}^{\infty} \frac{\sin(\omega T)}{\omega} e^{i\omega t} d\omega = A$$

provided $|t| \leq T$. Let $t = 0$ and we get

$$\frac{1}{\pi} \int_{-\infty}^{\infty} \frac{\sin(\omega T)}{\omega} d\omega = 1$$

Putting $T = 1$ and spotting that the integrand is even gives the result.

4. Using the definition of convolution given in the question, we have

$$f(at) * f(bt) = \int_{-\infty}^{\infty} f(a(t - \tau)) f(b\tau) d\tau.$$

$$= e^{-at} \int_{0}^{\infty} f(b\tau - a\tau) d\tau$$

$$= -e^{-at} \frac{1}{b - a} [f(bt - at) - 1]$$

$$= \frac{f(at) - f(bt)}{b - a}.$$

As $b \to a$ we have

$$f(at) * f(at) = -\lim_{b \to a} \frac{f(bt) - f(at)}{b - a}$$

$$= -\frac{d}{da} f(at) = -tf'(at) = tf(at).$$

5. With

$$g(x) = \int_{-\infty}^{\infty} f(t)e^{-2\pi i x t} dt$$

let $u = t - 1/2x$, so that $du = dt$. This gives

$$e^{-2\pi i x t} = e^{-2\pi i x(u+1/2x)} = -e^{-2\pi i x u}$$

Adding these two versions of $g(x)$ gives

$$|g(x)| = \left| \frac{1}{2} \int_{-\infty}^{\infty} (f(u) - f(u+1/2x))e^{-2\pi i x u} du \right|$$

$$\leq \frac{1}{2} \int_{-\infty}^{\infty} |f(u) - f(u+1/2x)| du$$

and as $x \to \infty$, the right hand side $\to 0$. Hence

$$\int_{-\infty}^{\infty} f(t) \cos(2\pi x t) dt \to 0 \quad \text{and} \quad \int_{-\infty}^{\infty} f(t) \sin(2\pi x t) dt \to 0$$

which illustrates the Riemann–Lebesgue lemma.

6. First of all we note that

$$G(it) = \int_{-\infty}^{\infty} g(\omega)e^{-i^2\omega t} d\omega = \int_{-\infty}^{\infty} g(\omega)e^{\omega t} d\omega$$

therefore

$$\int_{-\infty}^{\infty} f(t)G(it) dt = \int_{-\infty}^{\infty} f(t) \int_{-\infty}^{\infty} g(\omega)e^{\omega t} d\omega dt.$$

Assuming that the integrals are uniformly convergent so that their order can be interchanged, the right hand side can be written

$$\int_{-\infty}^{\infty} g(\omega) \int_{-\infty}^{\infty} f(t)e^{\omega t} dt d\omega,$$

which is, straight away, in the required form

$$\int_{-\infty}^{\infty} g(\omega)F(i\omega) d\omega.$$

Changing the dummy variable from ω to t completes the proof.

7. Putting $f(t) = 1 - t^2$ where $f(t)$ is zero outside the range $-1 \leq t \leq 1$ and $g(t) = e^{-t}$, $0 \leq t < \infty$, we have

$$F(\omega) = \int_{-1}^{1} (1 - t^2)e^{-i\omega t} dt$$

and

$$G(\omega) = \int_0^\infty e^{-t} e^{i\omega t} dt.$$

Evaluating these integrals (the first involves integrating by parts twice!) gives

$$F(\omega) = \frac{4}{\omega^3} \left(\omega \cos(\omega) - \sin(\omega) \right)$$

and

$$G(\omega) = \frac{1}{1 + i\omega}.$$

Thus, using Parseval's formula from the previous question, the imaginary unit disappears and we are left with

$$\int_{-1}^1 (1+t)dt = \int_0^\infty \frac{4e^{-t}}{t^3} \left(t \cosh(t) - \sinh(t) \right) dt$$

from which the desired integral is 2. Using Parseval's theorem

$$\int_{-\infty}^\infty |f(t)|^2 dt = \frac{1}{2\pi} \int_{-\infty}^\infty |F(\omega)|^2 d\omega$$

we have that

$$\int_{-1}^1 (1 - t^2)^2 dt = \frac{1}{2\pi} \int_0^\infty \frac{16 \left(t \cos(t) - \sin(t) \right)^2}{t^6} dt.$$

Evaluating the integral on the left we get

$$\int_0^\infty \frac{\left(t \cos(t) - \sin(t) \right)^2}{t^6} dt = \frac{\pi}{15}.$$

8. The ordinary differential equation obeyed by the Laplace Transform $\bar{u}(x, s)$ is

$$\frac{d^2 \bar{u}(x, s)}{dx^2} - \frac{s}{k} \bar{u}(x, s) = -\frac{g(x)}{k}.$$

Taking the Fourier Transform of this we obtain the solution

$$\bar{u}(x, s) = \frac{1}{2\pi} \int_{-\infty}^\infty \frac{G(\omega)}{s + \omega^2 k} e^{i\omega x} d\omega$$

where

$$G(\omega) = \int_{-\infty}^\infty g(x) e^{-i\omega x} dx$$

is the Fourier Transform of $g(x)$. Now it is possible to write down the solution by inverting $\bar{u}(x, s)$ as the Laplace variable s only occurs in the denominator of a simple fraction. Inverting using standard forms thus gives

$$u(x, t) = \frac{1}{2\pi} \int_{-\infty}^\infty G(\omega) e^{i\omega} e^{-\omega^2 k t} d\omega.$$

It is possible by completing the square and using the kind of "tricks" seen in Section 3.2 (page 52 etc.) to convert this into the solution that can be obtained directly by Laplace Transforms and convolution, namely

$$u(x,t) = \frac{1}{2\sqrt{\pi t}} \int_{-\infty}^{\infty} e^{-(x-\tau)^2/4t} g(\tau) d\tau.$$

9. To convert the partial differential equation into the integral form is a straightforward application of the theory of Section 6.4. Taking Fourier transforms in x and y using the notation

$$v(\lambda, y) = \int_{-\infty}^{\infty} u(x,y) e^{-i\lambda x} dx$$

and

$$w(\lambda, \mu) = \int_{-\infty}^{\infty} v(\lambda, y) e^{-i\mu y} dy$$

we obtain

$$-\lambda^2 w - \mu^2 w + k^2 w = \int_{-\infty}^{\infty} \int_{-\infty}^{\infty} f(\xi, \eta) e^{-\lambda \xi} e^{-\mu \eta} d\xi d\eta.$$

Using the inverse Fourier Transform gives the answer in the question. The conditions are that for both $u(x,y)$ and $f(x,y)$ all first partial derivatives must vanish at $\pm \infty$.

10. The Fourier series written in complex form is

$$f(x) \sim \sum_{n=-\infty}^{\infty} c_n e^{inx}$$

where

$$c_n = \frac{1}{2\pi} \int_{-\pi}^{\pi} f(x) e^{-inx} dx.$$

Now it is straightforward to use the window function $W(x)$ to write

$$c_n = \frac{1}{2} \int_{-\infty}^{\infty} W\left(\frac{x-\pi}{2\pi}\right) f(x) e^{-inx} dx.$$

11. The easiest way to see how discrete Fourier Transforms are derived is to consider a sampled time series as the original time series $f(t)$ multiplied by a function that picks out the discrete (sampled) values leaving all other values zero. This function is related to the Shah function (train of Dirac-δ functions) is not necessarily (but is usually) equally spaced. It is designed by using the window function $W(x)$ met in the last question. With such a function, taking the Fourier Transform results in the finite sum of the kind seen in the question. The inverse is a similar evaluation, noting that because of the discrete nature of the function, there is a division by the total number of data points.

12. Inserting the values $\{1, 2, 1\}$ into the series of the previous question, $N = 3$ and $T = 1$ so we get the three values

$$F_0 = 1 + 2 + 1 = 4; \quad F_1 = 1 + 2e^{-2\pi i/3} + e^{-4\pi i/3} = e^{-2\pi i/3};$$

and

$$F_2 = 1 + 2e^{-4\pi i/3} + e^{-8\pi i/3} = e^{-4\pi i/3}.$$

13. Using the exponential forms of sine and cosine the Fourier Transforms are immediately

$$\mathcal{F}\{\cos(\omega_0 t)\} = \frac{1}{2}(\delta(\omega - \omega_0) + \delta(\omega + \omega_0))$$

$$\mathcal{F}\{\sin(\omega_0 t)\} = \frac{1}{2i}(\delta(\omega - \omega_0) - \delta(\omega + \omega_0)).$$

Exercises 7.7

1. In all of these examples, the location of the pole is obvious, and the residue is best found by use of the formula

$$\lim_{z \to a}(z - a)f(z)$$

where $z = a$ is the location of the simple pole. In these answers, the location of the pole is followed after the semicolon by its residue. Where there is more than one pole, the answers are sequential, poles first followed by the corresponding residues.

(i) $z = -1$; 1,

(ii) $z = 1$; -1,

(iii) $z = 1, 3i, -3i$; $\frac{1}{2}, \frac{5}{12}(3 - i), \frac{5}{12}(3 + i)$,

(iv) $z = 0, -2, -1$; $\frac{3}{2}, -\frac{5}{2}, 1$,

(v) $z = 0$; 1,

(vi) $z = n\pi$ $(-1)^n n\pi$, n integer,

(vii) $z = n\pi$; $(-1)^n e^{n\pi}$, n integer.

2. As in the first example, the location of the poles is straightforward. The methods vary. For parts (i), (ii) and (iii) the formula for finding the residue at a pole of order n is best, viz.

$$\frac{1}{(n - 1)!} \lim_{z \to a} \frac{d^{(n-1)}}{dz^{(n-1)}}(z - a)^n f(z).$$

For part (iv) expanding both numerator and denominator as power series and picking out the coefficient of $1/z$ works best. The answers are as follows

(i) $z = 1$, order 2   res $= 4$

(ii) $z = i$, order 2   res $= -\frac{1}{4}i$

$z = -i$, order 2   res $= \frac{1}{4}i$

(iii) $z = 0$, order 3   res $= -\frac{1}{2}$

(iv) $z = 0$, order 2   res $= 1$.

3. (i) Using the residue theorem, the integral is $2\pi i$ times the sum of the residues of the integrand at the three poles. The three residues are:

$$\frac{1}{3}(1 - i) \text{ (at } z = 1), \quad \frac{4}{15}(-2 - i) \text{ (at } z = -2), \quad \frac{1}{5}(1 + 3i) \text{ (at } z = -i).$$

The sum of these times $2\pi i$ gives the result

$$-\frac{2\pi}{15}.$$

(ii) This time the residue (calculated easily using the formula) is 6, whence the integral is $12\pi i$.

(iii) For this integral we use a semi circular contour on the upper half plane. By the estimation lemma, the integral around the curved portion tends to zero as the radius gets very large. Also the integral from $-\infty$ to 0 along the real axis is equal to the integral from 0 to ∞ since the integrand is even. Thus we have

$$2\int_0^\infty \frac{1}{x^6 + 1}dx = 2\pi i(\text{sum of residues at } z = e^{\pi i/6}, \ i, \ e^{5\pi i/6})$$

from which we get the answer $\frac{\pi}{3}$.

(iv) This integral is evaluated using the same contour, and similar arguments tell us that

$$2\int_0^\infty \frac{\cos(2\pi x)}{x^4 + x^2 + 1}dx = 2\pi i(\text{sum of residues at } z = e^{\pi i/3}, \ e^{2\pi i/3}).$$

(Note that the complex function considered is $\dfrac{e^{2\pi i z}}{z^4 + z^2 + 1}$. Note also that the poles of the integrand are those of $z^6 - 1$ but excluding $z = \pm 1$.) The answer is, after a little algebra

$$-\frac{\pi}{2\sqrt{3}}e^{-\pi/\sqrt{3}}.$$

4. Problems (i) and (ii) are done by using the function

$$f(z) = \frac{(\ln(z))^2}{z^4 + 1}.$$

Integrated around the indented semi circular contour of Figure 7.8, there are poles at $z = (\pm 1 \pm i)/\sqrt{2}$. Only those at $(\pm 1 + i)/\sqrt{2}$ or $z = e^{\pi i/4}$, $e^{3\pi i/4}$ are inside the contour. Evaluating

$$\int_{C'} f(z)dz$$

along all the parts of the contour gives the following contributions: those along the curved bits eventually contribute nothing (the denominator gets very large in absolute magnitude as the radius of the big semi-circle $\to$ ∞, the integral around the small circle $\to 0$ as its radius $r \to 0$ since $r(\ln r)^2 \to 0$.) The contributions along the real axis are

$$\int_0^\infty \frac{(\ln x)^2}{x^4 + 1} dx$$

along the positive real axis where $z = x$ and

$$\int_0^\infty \frac{(\ln x + i\pi)^2}{x^4 + 1} dx$$

along the negative real axis where $z = x e^{i\pi}$ so $\ln z = \ln x + i\pi$. The residue theorem thus gives

$$2 \int_0^\infty \frac{(\ln x)^2}{x^4 + 1} dx + 2\pi i \int_0^\infty \frac{\ln x}{x^4 + 1} dx - \pi^2 \int_0^\infty \frac{1}{x^4 + 1} dx$$
$$= 2\pi i \{\text{sum of residues}\}. \tag{A.1}$$

The residue at $z = a$ is given by

$$\frac{(\ln a)^2}{4a^3}$$

using the formula for the residue of a simple pole. These sum to

$$-\frac{\pi^2}{64\sqrt{2}}(8 - 10i).$$

Equating real and imaginary parts of Equation A.1 gives the answers

(i) $\displaystyle\int_0^\infty \frac{(\ln x)^2}{x^4 + 1} dx = \frac{3\pi^3 \sqrt{2}}{64}$; (ii) $\displaystyle\int_0^\infty \frac{\ln x}{x^4 + 1} dx = -\frac{\pi^2}{16}\sqrt{2}$

once the result

$$\int_0^\infty \frac{1}{x^4 + 1} dx = \frac{\pi}{2\sqrt{2}}$$

from Example 7.5(ii) is used.

(iii) The third integral also uses the indented semi circular contour of Figure 7.8. The contributions from the large and small semi circles are ultimately zero. There is a pole at $z = i$ which has residue $e^{\pi\lambda i/2}/2i$ and the straight parts contribute

$$\int_0^\infty \frac{x^\lambda}{1 + x^2} dx$$

(positive real axis), and

$$-\int_\infty^0 \frac{x^\lambda e^{\lambda i\pi}}{1 + x^2} dx$$

(negative real axis). Putting the contributions together yields

$$\int_0^\infty \frac{x^\lambda}{1 + x^2} dx + e^{\lambda i\pi} \int_0^\infty \frac{x^\lambda}{1 + x^2} dx = \pi e^{\lambda i\pi/2}$$

from which

$$\int_0^\infty \frac{x^\lambda}{1 + x^2} dx = \frac{\pi}{2\cos(\frac{\lambda\pi}{2})}.$$

5. These inverse Laplace Transforms are all evaluated form first principles using the Bromwich contour, although it is possible to deduce some of them by using previously derived results, for example if we assume that

$$\mathcal{L}^{-1}\left\{\frac{1}{\sqrt{s}}\right\} = \frac{1}{\sqrt{\pi t}}$$

then we can carry on using the first shift theorem and convolution. However, we choose to use the Bromwich contour. The first two parts follow closely Example 7.10, though none of the branch points in these problems is in the exponential. The principle is the same.

(i) This Bromwich contour has a cut along the negative real axis from -1 to $-\infty$. It is shown as Figure A.6. Hence

$$\mathcal{L}^{-1}\left\{\frac{1}{s\sqrt{s + 1}}\right\} = \frac{1}{2\pi i}\int_{Br} \frac{e^{st}}{s\sqrt{s + 1}} ds.$$

The integral is thus split into the following parts

$$\int_{C'} = \int_{Br} + \int_\Gamma + \int_{AB} + \int_\gamma + \int_{CD} = 2\pi i(\text{residue at } s = 0)$$

where C' is the whole contour, Γ is the outer curved part, AB is the straight portion above the cut ($\Im s > 0$) γ is the small circle surrounding the branch point $s = -1$ and CD is the straight portion below the cut ($\Im s < 0$). The residue is one, the curved parts of the contour contribute nothing in the limit. The important contributions come from the integrals

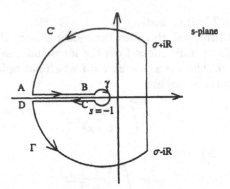

Figure A.6: The cut Bromwich contour

along AB and CD. On AB we can put $s = xe^{i\pi} - 1$. This leads to the integral

$$\int_{AB} = \int_{\infty}^{0} \frac{e^{t(-x-1)}}{(-x-1)i\sqrt{x}}dx$$

On CD we can put $s = xe^{-i\pi} - 1$. The integrals do not cancel because of the square root in the denominator (the reason the cut is there of course!). They in fact exactly reinforce. So the integral is

$$\int_{CD} = \int_{0}^{\infty} \frac{e^{t(-x-1)}}{(-x-1)(-i\sqrt{x})}dx.$$

Hence

$$\int_{C} = \int_{Br} -2\int_{\infty}^{0} \frac{e^{-t}e^{-xt}}{i(x+1)\sqrt{x}}dx = 2\pi i.$$

Using the integral

$$\int_{0}^{\infty} \frac{e^{-xt}}{(x+1)\sqrt{x}}dx = -\pi e^{t}[-1 + \text{erf}\sqrt{t}]$$

gives the answer that

$$\mathcal{L}^{-1}\left\{\frac{1}{s\sqrt{s+1}}\right\} = \frac{1}{2\pi i}\int_{Br} \frac{e^{st}}{s\sqrt{s+1}}ds = \text{erf}\sqrt{t}.$$

(ii) This second part is tackled in a similar way. The contour is identical (Figure A.6). The important step is the correct parametrisation of the straight parts of the contour just above and just below the branch cut. This time the two integrals along the cut are

$$\int_{AB} = -\int_{\infty}^{0} \frac{e^{(-x-1)t}}{1+i\sqrt{x}}dx$$

and

$$\int_{CD} = -\int_0^\infty \frac{e^{(-x-1)t}}{1 - i\sqrt{x}} dx,$$

and the integrals combine to give

$$\mathcal{L}^{-1}\left\{\frac{1}{1 + \sqrt{s+1}}\right\} = \frac{1}{2\pi i}\int_0^\infty \frac{e^{-t}e^{-xt}2i\sqrt{x}}{1 + x} dx$$

which gives the result

$$\frac{e^{-t}}{\sqrt{\pi t}} - \text{erfc}\sqrt{t}.$$

(iii) This inverse can be obtained from part (ii) by using the first shift theorem (Theorem 1.3, p10). The result is

$$\mathcal{L}^{-1}\left\{\frac{1}{1 + \sqrt{s}}\right\} = \frac{1}{\sqrt{\pi t}} - e^t \text{erfc}\sqrt{t}.$$

(iv) Finally this last part can be deduced by using the formula

$$\frac{1}{\sqrt{s}+1} - \frac{1}{\sqrt{s}-1} = -2\frac{1}{s-1}$$

The answer is

$$\mathcal{L}^{-1}\left\{\frac{1}{\sqrt{s}-1}\right\} = \frac{1}{\sqrt{\pi t}} + e^t(1 + \text{erf}\sqrt{t}).$$

6. This problem is best tackled by use of power series, especially so as there is no problem manipulating them as these exponential and related functions have series that are uniformly convergent for all finite values of t and s, excluding $s = 0$. The power series for the error function (obtained by integrating the exponential series term by term) yields:

$$\text{erf}\left(\frac{1}{s}\right) = \frac{2}{\sqrt{\pi}}\sum_{n=0}^\infty \frac{(-1)^n}{s^{2n+1}(2n+1)n!}.$$

Taking the inverse Laplace Transform using linearity and standard forms gives

$$\phi(t) = \frac{2}{\sqrt{\pi}}\sum_{n=0}^\infty \frac{(-1)^n}{(2n+1)n!}\frac{t^{2n}}{(2n)!}.$$

After some tidying up, this implies that

$$\phi(\sqrt{t}) = \frac{2}{\sqrt{\pi}}\sum_{n=0}^\infty \frac{(-1)^n t^n}{(2n+1)!n!}.$$

Taking the Laplace Transform of this series term by term gives

$$\mathcal{L}\left\{\phi(\sqrt{t})\right\} = \frac{2}{\sqrt{\pi s}} \sum_{n=0}^{\infty} (-1)^n \frac{(1/\sqrt{s})^{2n+1}}{(2n+1)!}$$

which is

$$\mathcal{L}\left\{\phi(\sqrt{t})\right\} = \frac{2}{\sqrt{\pi s}} \sin\left(\frac{1}{\sqrt{s}}\right)$$

as required.

7. This problem is tackled in a very similar way to the previous one. We simply integrate the series term by term and have to recognise

$$\sum_{k=0}^{\infty} \frac{(-1)^k (2k)!}{(k!)^2} \left(\frac{x}{s}\right)^{2k}$$

as the binomial series for

$$\left(1 + \frac{x^2}{s^2}\right)^{-1/2}.$$

Again, the series are uniformly convergent except for $s = \pm ix$ which must be excluded alongside $s = 0$.

8. The integrand

$$\frac{\cosh(x\sqrt{s})}{s \cosh(\sqrt{s})}$$

has a singularity at the origin and wherever $\sqrt{s}$ is an odd multiple of $\pi/2$. The presence of the square roots leads one to expect branch points, but in fact there are only simple poles. There are however infinitely many of them at locations

$$s = -\left(n + \frac{1}{2}\right)^2 \pi^2$$

and at the origin. The (uncut) Bromwich contour can thus be used; all the singularities are certainly to the left of the line $s = \sigma$ in Figure 7.9. The inverse is thus

$$\frac{1}{2\pi i} \int_{Br} e^{st} \frac{\cosh(x\sqrt{s})}{s \cosh(\sqrt{s})} ds = \text{sum of residues.}$$

The residue at $s = 0$ is straightforwardly

$$\lim_{s \to 0} (s - 0) \left\{ \frac{e^{st} \cosh(x\sqrt{s})}{s \cosh(\sqrt{s})} \right\} = 1.$$

The residue at the other poles is also calculated using the formula, but the calculation is messier, and the result is

$$\frac{4(-1)^n}{\pi(2n-1)} e^{-(n-1/2)^2 \pi^2 t} \cos\left(n - \frac{1}{2}\right) \pi x.$$

Thus we have

$$\mathcal{L}^{-1}\left\{\frac{\cosh(x\sqrt{s})}{s\cosh(\sqrt{s})}\right\} = 1 + \frac{4}{\pi}\sum_{n=1}^{\infty}\frac{(-1)^n}{2n-1}e^{-(n-\frac{1}{2})^2\pi^2 t}\cos\left(n-\frac{1}{2}\right)\pi x.$$

9. The Bromwich contour for the function

$$e^{-s^{\frac{1}{3}}}$$

has a cut from the branch point at the origin. We can thus use the contour depicted in Figure 7.10. As the origin has been excluded the integrand

$$e^{st-s^{\frac{1}{3}}}$$

has no singularities in the contour, so by Cauchy's theorem the integral around C' is zero. As is usual, the two parts of the integral that are curved give zero contribution as the outer radius of Γ gets larger and larger, and inner radius of the circle γ gets smaller. This is because $\cos\theta < 0$ on the left of the imaginary axis which means that the exponent in the integrand is negative on Γ, also on γ the ds contributes a zero as the radius of the circle γ decreases. The remaining contributions are

$$\int_{AB} = -\int_{\infty}^{0}e^{-xt-x^{\frac{1}{3}}e^{i\pi/3}}dx$$

and

$$\int_{CD} = -\int_{0}^{\infty}e^{-xt-x^{\frac{1}{3}}e^{-i\pi/3}}dx.$$

These combine to give

$$\int_{AB} + \int_{CD} = -\int_{0}^{\infty}e^{-xt-\frac{1}{2}x^{\frac{1}{3}}}\sin\left(\frac{x^{\frac{1}{3}}\sqrt{3}}{2}\right)dx.$$

Substituting $x = u^3$ gives the result

$$\mathcal{L}^{-1}\{e^{-s^{\frac{1}{3}}}\} = \frac{1}{2\pi i}\int_{Br}e^{st-s^{\frac{1}{3}}}ds = \frac{3}{\pi}\int_{0}^{\infty}u^2 e^{-u^3 t-\frac{1}{2}u}\sin\left(\frac{u\sqrt{3}}{2}\right)du.$$

10. Using Laplace Transforms in t solving in the usual way gives the solution

$$\bar{\phi}(x,s) = \frac{1}{s^2}e^{-x\sqrt{s^2-1}}.$$

The singularity of $\bar{\phi}(x,s)$ with the largest real part is at $s = 1$. The others are at $s = -1, 0$. Expanding $\bar{\phi}(x,s)$ about $s = 1$ gives

$$\bar{\phi}(x,s) = 1 - x\sqrt{2}(s-1)^{\frac{1}{2}} + \cdots.$$

In terms of Theorem 7.4 this means that $k = 1/2$ and $a_0 = -x\sqrt{2}$. Hence the leading term in the asymptotic expansion for $\phi(x, t)$ for large t is

$$-\frac{1}{\pi}e^t \sin\left(\frac{1}{2}\pi\right)\left(-x\sqrt{2}\frac{\Gamma(3/2)}{t^{3/2}}\right)$$

whence

$$\phi(x, t) \sim \frac{xe^t}{\sqrt{2\pi t^3}}$$

as required.

Table of Laplace Transforms

In this table, t is a real variable, s is a complex variable and a and b are real constants. In a few entries, the real variable x also appears.

$$f(s)\left(-\int_0^\infty e^{-st} F(t)\,dt\right) \qquad F(t)$$

$f(s)\left(-\int_0^\infty e^{-st}F(t)dt\right)$	$F(t)$
$\dfrac{1}{s}$	1
$\dfrac{1}{s^n}, n = 1, 2, \ldots$	$\dfrac{t^{n-1}}{(n-1)!},$
$\dfrac{1}{s^x}, x > 0,$	$\dfrac{t^{x-1}}{\Gamma(x)},$
$\dfrac{1}{s-a},$	e^{at}
$\dfrac{s}{s^2 + a^2}$	$\cos(at)$
$\dfrac{a}{s^2 + a^2}$	$\sin(at)$
$\dfrac{s}{s^2 - a^2}$	$\cosh(at)$
$\dfrac{a}{s^2 - a^2}$	$\sinh(at)$

1	$\delta(t)$
s^n	$\delta^{(n)}(t)$
$\dfrac{e^{-as}}{s}$	$H(t-a)$
$\dfrac{e^{-a\sqrt{s}}}{s}$	$\text{erfc}\left(\dfrac{a}{2\sqrt{t}}\right)$
$e^{-a\sqrt{s}}$	$\dfrac{a}{2\sqrt{\pi t^3}}e^{-a^2/4t}$
$\dfrac{1}{s\sqrt{s+a}}$	$\dfrac{\text{erf}\sqrt{at}}{\sqrt{a}}$
$\dfrac{1}{\sqrt{s}(s-a)}$	$\dfrac{e^{at}\text{erf}\sqrt{at}}{\sqrt{a}}$
$\dfrac{1}{\sqrt{s-a}+b}$	$e^{at}\left\{\dfrac{1}{\sqrt{\pi t}}-be^{b^2t}\text{erfc}(b\sqrt{t})\right\}$
$\dfrac{1}{\sqrt{s^2+a^2}}$	$J_0(at)$
$\tan^{-1}(a/s)$	$\dfrac{\sin(at)}{t}$
$\dfrac{\sinh(sx)}{s\sinh(sa)}$	$\dfrac{x}{a}+\dfrac{2}{\pi}\displaystyle\sum_{n=1}^{\infty}\dfrac{(-1)^n}{n}\sin\left(\dfrac{n\pi x}{a}\right)\cos\left(\dfrac{n\pi t}{a}\right)$
$\dfrac{\sinh(sx)}{s\cosh(sa)}$	$\dfrac{4}{\pi}\displaystyle\sum_{n=1}^{\infty}\dfrac{(-1)^n}{2n-1}\sin\left(\dfrac{(2n-1)\pi x}{2a}\right)\sin\left(\dfrac{(2n-1)\pi t}{2a}\right)$
$\dfrac{\cosh(sx)}{s\cosh(sa)}$	$1+\dfrac{4}{\pi}\displaystyle\sum_{n=1}^{\infty}\dfrac{(-1)^n}{2n-1}\cos\left(\dfrac{(2n-1)\pi x}{2a}\right)\cos\left(\dfrac{(2n-1)\pi t}{2a}\right)$
$\dfrac{\cosh(sx)}{s\sinh(sa)}$	$\dfrac{t}{a}+\dfrac{2}{\pi}\displaystyle\sum_{n=1}^{\infty}\dfrac{(-1)^n}{n}\cos\left(\dfrac{n\pi x}{a}\right)\sin\left(\dfrac{n\pi t}{a}\right)$
$\dfrac{\sinh(x\sqrt{s})}{s\sinh(a\sqrt{s})}$	$\dfrac{x}{a}+\dfrac{2}{\pi}\displaystyle\sum_{n=1}^{\infty}\dfrac{(-1)^n}{n}e^{-n^2\pi^2t/a^2}\sin\left(\dfrac{n\pi x}{a}\right)$

$$\frac{\cosh(x\sqrt{s})}{s\cosh(a\sqrt{s})}$$

$$1 + \frac{4}{\pi}\sum_{n=1}^{\infty}\frac{(-1)^n}{2n-1}e^{-(2n-1)^2\pi^2t/4a^2}\cos\left(\frac{(2n-1)\pi x}{2a}\right)$$

$$\frac{\sinh(x\sqrt{s})}{s^2\sinh(a\sqrt{s})}$$

$$\frac{xt}{a} + \frac{2a^2}{\pi^3}\sum_{n=1}^{\infty}\frac{(-1)^n}{n^3}(1 - e^{-n^2\pi^2t/a^2})\sin\left(\frac{n\pi x}{a}\right)$$

$$\frac{\sinh(sx)}{s^2\cosh(sa)}$$

$$x + \frac{8a}{\pi^2}\sum_{n=1}^{\infty}\frac{(-1)^n}{(2n-1)^2}\sin\left(\frac{(2n-1)\pi x}{2a}\right)\cos\left(\frac{(2n-1)\pi t}{2a}\right)$$

$$\frac{1}{as^2}\tanh\left(\frac{as}{2}\right)$$

$$F(t) = \begin{cases} t/a & 0 \le t \le a \\ 2 - t/a & a < t \le 2a \end{cases} \quad F(t) = F(t+2a)$$

$$\frac{1}{s}\tanh\left(\frac{as}{2}\right)$$

$$F(t) = \begin{cases} 1 & 0 \le t \le a \\ -1 & a < t \le 2a \end{cases} \quad F(t) = F(t+2a)$$

$$\frac{\pi a}{a^2 + s^2}\coth\left(\frac{as}{2}\right)$$

$$\left|\sin\left(\frac{\pi t}{a}\right)\right|$$

$$\frac{\pi a}{(a^2s^2 + \pi^2)(1 - e^{-as})}$$

$$F(t) = \begin{cases} \sin(\frac{\pi t}{a}) & 0 \le t \le a \\ 0 & a < t \le 2a \end{cases} \quad F(t) = F(t+2a)$$

$$\frac{1}{as^2} - \frac{e^{-as}}{s(1 - e^{-as})}$$

$$F(t) = t/a, \quad 0 \le t \le a \quad F(t) = F(t+a)$$

C
Linear Spaces

C.1 Linear Algebra

In this appendix, some fundamental concepts of linear algebra are given. The proofs are largely omitted; students are directed to textbooks on linear algebra for these. For this subject, we need to be precise in terms of the basic mathematical notions and notations we use. Therefore we uncharacteristically employ a formal mathematical style of prose. It is essential to be rigorous with the basic mathematics, but it is often the case that an over formal treatment can obscure rather than enlighten. That is why this material appears in an appendix rather than in the main body of the text.

A set of objects (called the elements of the set) is written as a sequence of (usually) lower case letters in between braces:-

$$A = \{a_1, a_2, a_3, \ldots, a_n\}.$$

In discussing Fourier series, sets have infinitely many elements, so there is a row of dots after a_n too. The symbol $\in$ read as "belongs to" should be familiar to most. So $s \in S$ means that s is a member of the set S. Sometimes the alternative notation

$$S = \{x | f(x)\}$$

is used. The vertical line is read as "such that" so that $f(x)$ describes some property that x possess in order that $s \in S$. An example might be

$$S = \{x | x \in \mathbf{R}, |x| \leq 2\}$$

so S is the set of real numbers that lie between -2 and $+2$.

The following notation is standard but is reiterated here for reference:

(a, b) denotes the open interval $\{x | a < x < b\}$,

$[a, b]$ denotes the closed interval $\{x | a \le x \le b\}$,

$[a, b)$ is the set $\{x | a \le x < b\}$,

and $(a, b]$ is the set $\{x | a < x \le b\}$.

The last two are described as half closed or half open intervals and are reasonably obvious extensions of the first two definitions. In Fourier series, ∞ is often involved so the following intervals occur:-

$$(a, \infty) = \{x | a < x\} \quad [a, \infty) = \{x | a \le x\}$$
$$(-\infty, a) = \{x | x < a\} \quad (-\infty, a] = \{x | x \le a\}.$$

These are all obvious extensions. Where appropriate, use is also made of the following standard sets

$\mathbf{Z} = \{\ldots, -3, -2, -1, 0, 1, 2, 3, \ldots\}$, the set of integers

$\mathbf{Z}_+$ is the set of positive integers including zero: $\{0, 1, 2, 3, \ldots\}$

$\mathbf{R} = \{x | x \text{ is a real number}\} = (-\infty, \infty)$.

Sometimes (but very rarely) we might use the set of fractions or rationals $\mathbf{Q}$:

$$\mathbf{Q} = \left\{ \frac{m}{n} \mid m \text{ and } n \text{ are integers}, n \ne 0 \right\}.$$

$\mathbf{R}_+$ is the set of positive real numbers

$$\mathbf{R}_+ = \{x \mid x \in \mathbf{R}, x \ge 0\}.$$

Finally the set of complex numbers $\mathbf{C}$ is defined by

$$\mathbf{C} = \{z = x + iy \mid x, y \in \mathbf{R}, i = \sqrt{-1}\}.$$

The standard notation

$$x = \Re\{z\}, \quad \text{the real part of } z$$
$$y = \Im\{z\}, \quad \text{the imaginary part of } z$$

has already been met in Chapter 1.

Hopefully, all of this is familiar to most of you. We will need these to define the particular normed spaces within which Fourier series operate. This we now proceed to do. A vector space V is an algebraic structure that consists of elements (called vectors) and two operations (called addition + and multiplication ×). The following gives a list of properties obeyed by vectors $\mathbf{a}, \mathbf{b}, \mathbf{c}$ and scalars $\alpha, \beta \in F$ where F is a field (usually $\mathbf{R}$ or $\mathbf{C}$).

1. $\mathbf{a} + \mathbf{b}$ is also a vector (closure under addition).

2. $(\mathbf{a} + \mathbf{b}) + \mathbf{c} = \mathbf{a} + (\mathbf{b} + \mathbf{c})$ (associativity under addition).

3. There exists a zero vector denoted by 0 such that $0 + a = a + 0 = a \ \forall a \in V$ (additive identity).

4. For every vector $a \in V$ there is a vector $-a$ (called "minus a" such that $a + (-a) = 0$.

5. $s + b = b + a$ for every $a, b \in V$ (additive commutativity).

6. $\alpha a \in V$ for every $\alpha \in F, a \in V$ (scalar multiplicity).

7. $\alpha(a + b) = \alpha a + \alpha b$ for every $\alpha \in F, a, b \in V$ (first distributive law).

8. $(\alpha + \beta)a = \alpha a + \beta a$ for every $\alpha, \beta \in F, a \in V$ (second distributive law).

9. For the unit scalar 1 of the field F, and every $a \in V$ $1.a = a$ (multiplicative identity).

The set V whose elements obey the above nine properties over a field F is called a vector space over F. The name linear space is also used in place of vector space and is useful as the name "vector" conjures up mechanics to many and gives a false impression in the present context. In the study of Fourier series, vectors are in fact functions. The name linear space emphasises the linearity property which is confirmed by the following definition and properties.

Definition C.1 *If* $a_1, a_2, \ldots, a_n \in V$ *where V is a linear space over a field F and if there exist scalars* $\alpha_1, \alpha_2, \ldots, \alpha_n \in F$ *such that*

$$b = \alpha_1 a_1 + \alpha_2 a_2 + \cdots + \alpha_n a_n$$

(called a linear combination *of the vectors* $a_1, a_2, \ldots, a_n$*) then the collection of all such* b *which are a linear combination of the vectors* $a_1, a_2, \ldots, a_n$ *is called the* span *of* $a_1, a_2, \ldots, a_n$ *denoted by* $\mathrm{span}\{a_1, a_2, \ldots, a_n\}$.

If this definition seems innocent, then the following one which depends on it is not. It is one of the most crucial properties possibly in the whole of mathematics.

Definition C.2 (linear independence) *If V is a linear (vector) space, the vectors* $a_1, a_2, \ldots, a_n \in V$ *are said to be linearly independent if the equation*

$$\alpha_1 a_1 + \alpha_2 a_2 + \cdots + \alpha_n a_n = 0$$

implies that all of the scalars are zero, i.e.

$$\alpha_1 = \alpha_2 = \ldots = \alpha_n = 0, (\alpha_1, \alpha_2, \ldots, \alpha_n \in F).$$

Otherwise, $\alpha_1, \alpha_2, \ldots, \alpha_n$ *are said to be linearly dependent.*

Again, it is hoped that this is not a new concept. However, here is an example. Most texts take examples from geometry, true vectors indeed. This is not appropriate here so instead this example is algebraic.

Example C.3 *Is the set* $S = \{1, x, 2+x, x^2\}$ *with* $F = \mathbf{R}$ *linearly independent?*

Solution The most general combination of $1, x, 2+x, x^2$ is

$$y = \alpha_1 + \alpha_2 x + \alpha_3(2+x) + \alpha_4 x^2$$

where x is a variable that can take any real value.

Now, $y = 0$ for all x does not imply $\alpha_1 = \alpha_2 = \alpha_3 = \alpha_4 = 0$, for if we choose $\alpha_1 + 2\alpha_2 = 0, \alpha_2 + \alpha_3 = 0$ and $\alpha_4 = 0$ then $y = 0$. The combination $\alpha_1 = 1, \alpha_3 = -\frac{1}{2}, \alpha_2 = \frac{1}{2}, \alpha_4 = 0$ will do. The set is therefore not linearly independent.

On the other hand, the set $\{1, x, x^2\}$ is most definitely linearly independent as

$$\alpha_1 + \alpha_2 x + \alpha_3 x^2 = 0 \text{ for all } x \Rightarrow \alpha_1 = \alpha_2 = \alpha_3 = 0.$$

It is possible to find many independent sets. One could choose $\{x, \sin(x), \ln x\}$ for example: however sets like this are not very useful as they do not lead to any applications. The set $\{1, x, x^2\}$ spans all quadratic functions. Here is another definition that we hope is familiar.

Definition C.4 *A finite set of vectors* $\mathbf{a_1, a_2, \ldots, a_n}$ *is said to be a* basis *for the linear space* V *if the set of vectors* $\mathbf{a_1, a_2, \ldots, a_n}$ *is linearly independent and* $V = span\{\mathbf{a_1, a_2, \ldots, a_n}\}$ *The natural number* n *is called the dimension of* V *and we write* $n = dim(V)$.

Example C.5 *Let* $[a, b]$ *(with* $a < b$) *denote the finite closed interval as already defined. Let* f *be a continuous real valued function whose value at the point* x *of* $[a, b]$ *is* $f(x)$. *Let* $C[a, b]$ *denote the set of all such functions. Now, if we define addition and scalar multiplication in the natural way, i.e.* $f_1 + f_2$ *is simply the value of* $f_1(x) + f_2(x)$ *and similarly* αf *is the value of* $\alpha f(x)$, *then it is clear that* $C[a, b]$ *is a real vector space. In this case, it is clear that the set* $x, x^2, x^3, \ldots, x^n$ *are all members of* $C[a, b]$ *for arbitrarily large* n. *It is therefore not possible for* $C[a, b]$ *to be finite dimensional.*

Perhaps it is now a little clearer as to why the set $\{1, x, x^2\}$ is useful as this is a basis for all quadratics, whereas $\{x, \sin(x), \ln x\}$ does not form a basis for any well known space. Of course, there is usually an infinite choice of basis for any particular linear space. For the quadratic functions the sets $\{1 - x, 1 + x, x^2\}$ or $\{1, 1 - x^2, 1 + 2x + x^2\}$ will do just as well. That we have these choices of bases is useful and will be exploited later.

Most books on elementary linear algebra are content to stop at this point and consolidate the above definitions through examples and exercises. However, we need a few more definitions and properties in order to meet the requirements of a Fourier series.

Definition C.6 *Let* V *be a real or complex linear space. (That is the field* F *over which the space is defined is either* $\mathbf{R}$ *or* $\mathbf{C}$.) *An inner product is an operation between two elements of* V *which results in a scalar. This scalar is denoted by* $\langle \mathbf{a_1, a_2} \rangle$ *and has the following properties:-*

1. *For each* $a_1 \in V$, $\langle a_1, a_1 \rangle$ *is a non-negative real number, i.e.*

$$\langle a_1, a_1 \rangle \geq 0.$$

2. *For each* $a_1 \in V$, $\langle a_1, a_1 \rangle = 0$ *if and only if* $a_1 = 0$.

3. *For each* $a_1, a_2, a_3 \in V$ *and* $\alpha_1, \alpha_2 \in F$

$$\langle \alpha_1 a_1 + \alpha_2 a_2, a_3 \rangle = \alpha_1 \langle a_1, a_3 \rangle + \alpha_2 \langle a_2, a_3 \rangle.$$

4. *For each* $a_1, a_2 \in V$, $\langle a_1, a_2 \rangle = \overline{\langle a_2, a_1 \rangle}$

where the overbar in the last property denotes the complex conjugate. If $F = \mathbf{R}$ α_1, α_2 *are real, and Property 4 becomes obvious.*

No doubt, students who are familiar with the geometry of vectors will be able to identify the inner product $\langle a_1, a_2 \rangle$ with $a_1.a_2$ the scalar product of the two vectors a_1 and a_2. This is one useful example, but it is by no means essential to the present text where most of the inner products take the form of integrals.

Inner products provide a rich source of properties that would be out of place to dwell on or prove here. For example:

$$\langle 0, a \rangle = 0 \quad \forall a \in V$$

and

$$\langle \alpha a_1, \alpha a_2 \rangle = |\alpha|^2 \langle a_1, a_2 \rangle.$$

Instead, we introduce two examples of inner product spaces.

1. If $\mathbf{C}^n$ is the vector space V, i.e. a typical element of V has the form $a = (a_1, a_2, \ldots, a_n)$ where $a_r = x_r + iy_r$, x_r, $y_r \in \mathbf{R}$. The inner product $\langle a, b \rangle$ is defined by

$$\langle a, b \rangle = a_1 \overline{b_1} + a_2 \overline{b_2} + \cdots + a_n \overline{b_n},$$

the overbar denoting complex conjugate.

2. Nearer to our applications of inner products is the choice $V = C[a, b]$ the linear space of all continuous functions f defined on the closed interval $[a, b]$. With the usual summation of functions and multiplication by scalars this can be verified to be a vector space over the field of complex numbers $\mathbf{C}^n$. Given a pair of continuous functions f, g we can define their inner product by

$$\langle f, g \rangle = \int_a^b f(x) \overline{g(x)} dx.$$

It is left to the reader to verify that this is indeed an inner product space satisfying the correct properties in Definition C6.

It is quite typical for linear spaces involving functions to be infinite dimensional. In fact it is very unusual for it to be otherwise!

What has been done so far is to define a linear space and an inner product on that space. It is nearly always true that we can define what is called a "norm" on a linear space. The norm is independent of the inner product in theory, but there is almost always a connection in practice. The norm is a generalisation of the notion of distance. If the linear space is simply two or three dimensional vectors, then the norm can indeed be distance. It is however, even in this case possible to define others. Here is the general definition of norm.

Definition C.7 *Let V be a linear space. A norm on V is a function from V to $\mathbf{R}_+$ (non-negative real numbers), denoted by being placed between two vertical lines $\|.\|$ which satisfies the following four criteria:-*

1. For each $\mathbf{a}_1 \in V$, $\|\mathbf{a}_1\| \geq 0$.

2. $\|\mathbf{a}_1\| = 0$ if and only if $\mathbf{a}_1 = \mathbf{0}$.

3. For each $\mathbf{a}_1 \in V$ and $\alpha \in \mathbf{C}$

$$\|\alpha \mathbf{a}_1\| = |\alpha| \|\mathbf{a}_1\|.$$

4. For every $\mathbf{a}_1, \mathbf{a}_2 \in V$

$$\|\mathbf{a}_1 + \mathbf{a}_2\| \leq \|\mathbf{a}_1\| + \|\mathbf{a}_2\|.$$

(4 is the triangle inequality.)

For the vector space comprising the elements $\mathbf{a} = (a_1, a_2, \ldots, a_n)$ where $a_r = x_r + iy_r$, $x_r, y_r \in \mathbf{R}$, i.e. $\mathbf{C}^n$ met previously, the obvious norm is

$$\begin{aligned}\|\mathbf{a}\| &= [|a_1|^2 + |a_2|^2 + |a_3|^2 + \cdots + |a_n|^2]^{1/2} \\ &= [\langle \mathbf{a}, \mathbf{a} \rangle]^{1/2}.\end{aligned}$$

It is true in general that we can always define the norm $\|.\|$ of a linear space equipped with an inner product $\langle ., . \rangle$ to be such that

$$\|\mathbf{a}\| = [\langle \mathbf{a}, \mathbf{a} \rangle]^{1/2}.$$

This norm is used in the next example. A linear space equipped with an inner product is called an inner product space. The norm induced by the inner product, sometimes called the *natural* norm for the function space $C[a, b]$, is

$$\|f\| = \left[\int_a^b |f|^2 dx \right]^{1/2}.$$

For applications to Fourier series we are able to make $\|f\| = 1$ and we adjust elements of V, i.e. $C[a, b]$ so that this is achieved. This process is called normalisation. Linear spaces with special norms and other properties are the directions in which this subject now naturally moves. The interested reader is directed towards books on functional analysis.

We now establish an important inequality called the Cauchy–Schwartz inequality. We state it in the form of a theorem and prove it.

Theorem C.8 (Cauchy–Schwartz) *Let V be a linear space with inner product $\langle .,. \rangle$, then for each $\mathbf{a}, \mathbf{b} \in V$ we have:*

$$|\langle \mathbf{a}, \mathbf{b} \rangle|^2 \leq ||\mathbf{a}||.||\mathbf{b}||.$$

Proof If $\langle \mathbf{a}, \mathbf{b} \rangle = 0$ then the result is self evident. We therefore assume that $\langle \mathbf{a}, \mathbf{b} \rangle = \alpha \neq 0$, α may of course be complex. We start with the inequality

$$||\mathbf{a} - \lambda \alpha \mathbf{b}||^2 \geq 0$$

where λ is a real number. Now,

$$||\mathbf{a} - \lambda \alpha \mathbf{b}||^2 = \langle \mathbf{a} - \lambda \alpha \mathbf{b}, \mathbf{a} - \lambda \alpha \mathbf{b} \rangle.$$

We use the properties of the inner product to expand the right hand side as follows:-

$$\langle \mathbf{a} - \lambda \alpha \mathbf{b}, \mathbf{a} - \lambda \alpha \mathbf{b} \rangle =$$
$$\langle \mathbf{a}, \mathbf{a} \rangle - \lambda \langle \alpha \mathbf{b}, \mathbf{a} \rangle - \lambda \langle \mathbf{a}, \alpha \mathbf{b} \rangle + \lambda^2 |\alpha|^2 \langle \mathbf{b}, \mathbf{b} \rangle \geq 0$$
$$\text{so } ||\mathbf{a}||^2 - \lambda \alpha \langle \mathbf{b}, \mathbf{a} \rangle - \lambda \bar{\alpha} \langle \mathbf{a}, \mathbf{b} \rangle + \lambda^2 |\alpha|^2 ||\mathbf{b}||^2 \geq 0$$
$$\text{i.e. } ||\mathbf{a}||^2 - \lambda \alpha \bar{\alpha} - \lambda \bar{\alpha} \alpha + \lambda^2 |\alpha|^2 ||\mathbf{b}||^2 \geq 0$$
$$\text{so } ||\mathbf{a}||^2 - 2\lambda |\alpha|^2 + \lambda^2 |\alpha|^2 ||\mathbf{b}||^2 \geq 0.$$

This last expression is a quadratic in the real parameter λ, and it has to be positive for all values of λ. The condition for the quadratic

$$a\lambda^2 + b\lambda + c$$

to be non-negative is that $b^2 \leq 4ac$ and $a > 0$. With

$$a = |\alpha|^2 ||\mathbf{b}||^2, \quad b = -2|\alpha|^2, \quad c = ||\mathbf{a}||^2$$

the inequality $b^2 \leq 4ac$ is

$$4|\alpha|^4 \leq 4|\alpha|^2 ||\mathbf{a}||^2 ||\mathbf{b}||^2$$
$$\text{or } |\alpha|^2 \leq ||\mathbf{a}|| ||\mathbf{b}||$$

and since $\alpha = \langle \mathbf{a}, \mathbf{b} \rangle$ the result follows.

$\square$

The following is an example that typifies the process of proving that something is a norm.

Example C.9 *Prove that* $||\mathbf{a}|| = \sqrt{(\mathbf{a}.\mathbf{a})} \in V$ *is indeed a norm for the vector space* V *with inner product* $\langle , \rangle$.

Proof The proof comprises showing that $\sqrt{\langle \mathbf{a}, \mathbf{a} \rangle}$ satisfies the four properties of a norm.

1. $||\mathbf{a}|| \geq 0$ follows immediately from the definition of square roots.

2. If $\mathbf{a} = \mathbf{0} \iff \sqrt{\langle \mathbf{a}, \mathbf{a} \rangle} = 0$.

3.

$$\begin{aligned}
||\alpha \mathbf{a}|| &= \sqrt{\langle \alpha \mathbf{a}, \alpha \mathbf{a} \rangle} = \\
\sqrt{\alpha \bar{\alpha} \langle \mathbf{a}, \mathbf{a} \rangle} &= \sqrt{|\alpha|^2 \langle \mathbf{a}, \mathbf{a} \rangle} = \\
|\alpha| \sqrt{\langle \mathbf{a}, \mathbf{a} \rangle} &= |\alpha| ||\mathbf{a}||.
\end{aligned}$$

4. This fourth property is the only one that takes a little effort to prove. Consider $||\mathbf{a} + \mathbf{b}||^2$. This is equal to

$$\begin{aligned}
\langle \mathbf{a} + \mathbf{b}, \mathbf{a} + \mathbf{b} \rangle &= \langle \mathbf{a}, \mathbf{a} \rangle + \langle \mathbf{b}, \mathbf{a} \rangle + \langle \mathbf{a}, \mathbf{b} \rangle + \langle \mathbf{b}, \mathbf{b} \rangle \\
&= ||\mathbf{a}||^2 + \langle \mathbf{a}, \mathbf{b} \rangle + \overline{\langle \mathbf{a}, \mathbf{b} \rangle} + ||\mathbf{b}||^2.
\end{aligned}$$

The expression $\langle \mathbf{a}, \mathbf{b} \rangle + \overline{\langle \mathbf{a}, \mathbf{b} \rangle}$, being the sum of a number and its complex conjugate, is real. In fact

$$\begin{aligned}
|\langle \mathbf{a}, \mathbf{b} \rangle + \overline{\langle \mathbf{a}, \mathbf{b} \rangle}| &= |2\Re \langle \mathbf{a}, \mathbf{b} \rangle| \\
&\leq 2|\langle \mathbf{a}, \mathbf{b} \rangle| \\
&\leq 2||\mathbf{a}|| . ||\mathbf{b}||
\end{aligned}$$

using the Cauchy–Schwartz inequality. Thus

$$\begin{aligned}
||\mathbf{a} + \mathbf{b}||^2 &= ||\mathbf{a}||^2 + \langle \mathbf{a}, \mathbf{b} \rangle + \overline{\langle \mathbf{a}, \mathbf{b} \rangle} + ||\mathbf{b}||^2 \\
&\leq ||\mathbf{a}||^2 + 2||\mathbf{a}|| ||\mathbf{b}|| + ||\mathbf{b}||^2 \\
&= (||\mathbf{a} + \mathbf{b}||)^2.
\end{aligned}$$

Hence $||\mathbf{a} + \mathbf{b}|| \leq ||\mathbf{a}|| + ||\mathbf{b}||$ which establishes the triangle inequality, Property 4.

Hence $||\mathbf{a}|| = \sqrt{\langle \mathbf{a}, \mathbf{a} \rangle}$ is a norm for V.

□

An important property associated with linear spaces is orthogonality. It is a direct analogy/generalisation of the geometric result $\mathbf{a.b} = 0, \mathbf{a} \neq \mathbf{0}, \mathbf{b} \neq \mathbf{0}$ if $\mathbf{a}$ and $\mathbf{b}$ represent directions that are at right angles (i.e. are orthogonal) to each other. This idea leads to the following pair of definitions.

Definition C.10 *Let V be an inner product space, and let $\mathbf{a}, \mathbf{b} \in V$. If $\langle \mathbf{a}, \mathbf{b} \rangle = 0$ then vectors $\mathbf{a}$ and $\mathbf{b}$ are said to be orthogonal.*

Definition C.11 *Let V be a linear space, and let $\{\mathbf{a}_1, \mathbf{a}_2, \ldots, \mathbf{a}_n\}$ be a sequence of vectors, $\mathbf{a}_r \in V$, $\mathbf{a}_r \neq 0$, $r = 1, 2, \ldots, n$, and let $\langle \mathbf{a}_i, \mathbf{a}_j \rangle = 0$, $i \neq j$, $0 \leq i, j \leq n$. Then $\{\mathbf{a}_1, \mathbf{a}_2, \ldots, \mathbf{a}_n\}$ is called an orthogonal set of vectors.*

Further to these definitions, a vector $\mathbf{a} \in V$ for which $\|\mathbf{a}\| = 1$ is called a unit vector and if an orthogonal set of vectors consists of all unit vectors, the set is called *orthonormal*.

It is also possible to let $n \to \infty$ and obtain an orthogonal set for an infinite dimensional inner product space. We make use of this later in this chapter, but for now let us look at an example.

Example C.12 *Determine an orthonormal set of vectors for the linear space that consists of all real linear functions:*

$$\{a + bx : a, b \in \mathbf{R} \ 0 \leq x \leq 1\}$$

using as inner product

$$\langle f, g \rangle = \int_0^1 f g \, dx.$$

Solution The set $\{1, x\}$ forms a basis, but it is not orthogonal. Let $a + bx$ and $c + dx$ be two vectors. In order to be orthogonal we must have

$$\langle a + bx, c + dx \rangle = \int_0^1 (a + bx)(c + dx) dx = 0.$$

Performing the elementary integration gives the following condition on the constants $a, b, c,$ and d

$$ac + \frac{1}{2}(bc + ad) + \frac{1}{3}bd = 0.$$

In order to be orthonormal too we also need

$$\|a + bx\| = 1 \text{ and } \|c + dx\| = 1$$

and these give, additionally,

$$a^2 + b^2 = 1, c^2 + d^2 = 1.$$

There are four unknowns and three equations here, so we can make a convenient choice. Let us set

$$a = -b = \frac{1}{\sqrt{2}}$$

which gives

$$\frac{1}{\sqrt{2}}(1-x)$$

as one vector. The first equation now gives $3c = -d$ from which

$$c = \frac{1}{\sqrt{10}}, \quad d = -\frac{3}{\sqrt{10}}.$$

Hence the set $\{(1-x)/\sqrt{10}, (1-3x)/\sqrt{10}\}$ is a possible orthonormal one.

Of course there are infinitely many possible orthonormal sets, the above was one simple choice. The next definition follows naturally.

Definition C.13 *In an inner product space, an orthonormal set that is also a basis is called an orthonormal basis.*

This next example involves trigonometry which at last gets us close to discussing Fourier series.

Example C.14 *Show that $\{\sin(x), \cos(x)\}$ is an orthogonal basis for the inner product space $V = \{a\sin(x) + b\cos(x) \,|\, a, b \in \mathbf{R}, 0 \leq x \leq \pi\}$ using as inner product*

$$\langle f, g \rangle = \int_0^1 fg\,dx, f, g \in V$$

and determine an orthonormal basis.

Solution V is two dimensional and the set $\{\sin(x), \cos(x)\}$ is obviously a basis. We merely need to check orthogonality. First of all,

$$\langle \sin(x), \cos(x) \rangle = \int_0^\pi \sin(x)\cos(x)dx = \frac{1}{2}\int_0^\pi \sin(2x)dx$$
$$= \left[-\frac{1}{4}\cos(2x)\right]_0^\pi$$
$$= 0.$$

Hence orthogonality is established. Also,

$$\langle \sin(x), \sin(x) \rangle = \int_0^\pi \sin^2(x)dx = \frac{\pi}{2}$$

and

$$\langle \cos(x), \cos(x) \rangle = \int_0^\pi \cos^2(x)dx = \frac{\pi}{2}.$$

Therefore

$$\left\{\sqrt{\frac{2}{\pi}}\sin(x), \sqrt{\frac{2}{\pi}}\cos(x)\right\}$$

is an orthonormal basis.

These two examples are reasonably simple, but for linear spaces of higher dimensions it is by no means obvious how to generate an orthonormal basis. One way of formally generating an orthonormal basis from an arbitrary basis is to use the Gramm–Schmidt orthonormalisation process, and this is given later in this appendix

There are some further points that need to be aired before we get to discussing Fourier series proper. These concern the properties of bases, especially regarding linear spaces of infinite dimension. If the basis $\{a_1, a_2, \ldots, a_n\}$ spans the linear space V, then *any* vector $v \in V$ can be expressed as a linear combination of the basis vectors in the form

$$v = \sum_{r=1}^{n} \alpha_r a_r.$$

This result follows from the linear independence of the basis vectors, and that they span V.

If the basis is orthonormal, then a typical coefficient, α_k can be determined by taking the inner product of the vector v with the corresponding basis vector a_k as follows

$$\langle v, a_k \rangle = \sum_{r=1}^{n} \alpha_r \langle a_r, a_k \rangle$$

$$= \sum_{r=1}^{n} \alpha_r \delta_{kr}$$

$$= \alpha_k$$

where δ_{kr} is Kronecker's delta:-

$$\delta_{kr} = \begin{cases} 1 & r = k \\ 0 & r \neq k \end{cases}.$$

If we try to generalise this to the case $n = \infty$ there are some difficulties. They are not insurmountable, but neither are they trivial. One extra need always arises when the case $n = \infty$ is considered and that is convergence. It is this that prevents the generalisation from being straightforward. The notion of *completeness* is also important. It has the following definition:

Definition C.15 *Let* $\{e_1, e_2, \ldots\ldots\}$ *be an infinite orthonormal system in an inner product space* V. *The system is* complete *in* V *if only the zero vector* $(u = 0)$ *satisfies the equation*

$$\langle u, e_n \rangle = 0, \quad n \in N$$

A complete inner product space whose basis has infinitely many elements is called a Hilbert Space, the properties of which take us beyond the scope of this short appendix on linear algebra. The next step would be to move on to Bessel's Inequality which is stated but not proved in Chapter 4.

Here are a few more definitions that help when we have to deal with series of vectors rather than series of scalars

Definition C.16 *Let* $w_1, w_2, \ldots, w_n, \ldots$ *be an infinite sequence of vectors in a normed linear space (e.g. an inner product space) W. We say that the sequence converges in norm to the vector* $w \in W$ *if*

$$\lim_{n \to \infty} \|w - w_n\| = 0.$$

This means that for each $\epsilon > 0$, there exists $n > n(\epsilon)$ such that $\|w - w_n\| < \epsilon, \forall n > n(\epsilon)$.

Definition C.17 *Let* $a_1, a_2, \ldots, a_n, \ldots$ *be an infinite sequence of vectors in the normed linear space V. We say that the series*

$$w_n = \sum_{r=1}^{n} \alpha_r a_r$$

converges in norm to the vector w if $\|w - w_n\| \to 0$ as $n \to \infty$. We then write

$$w = \sum_{r=1}^{\infty} \alpha_r a_r.$$

There is logic in this definition as $\|w_n - w\|$ measures the distance between the vectors w and w_n and if this gets smaller there is a sense in which w converges to w_n.

Definition C.18 *If $\{e_1, e_2, \ldots, e_n \ldots\}$ is an infinite sequence of orthonormal vectors in a linear space V we say that the system is closed in V if, for every $a \in V$ we have*

$$\lim_{n \to \infty} \|a - \sum_{r=1}^{n} \langle a, e_r \rangle e_r\| = 0.$$

There are many propositions that follow from these definitions, but for now we give one more definition that is useful in the context of Fourier series.

Definition C.19 *If $\{e_1, e_2, \ldots, e_n \ldots\}$ is an infinite sequence of orthonormal vectors in a linear space V of infinite dimension with an inner product, then we say that the system is complete in V if only the zero vector $a = 0$ satisfies the equation*

$$\langle a, e_n \rangle = 0, \quad n \in \mathbf{N}.$$

There are many more general results and theorems on linear spaces that are useful to call on from within the study of Fourier series. However, in a book such as this a judgement has to be made as when to stop the theory. Enough theory of linear spaces has now been covered to enable Fourier series to be put in proper context. The reader who wishes to know more about the pure mathematics of particular spaces can enable their thirst to be quenched by many excellent texts on special spaces. (Banach Spaces and Sobolev Spaces both have a special place in applied mathematics, so inputting these names in the appropriate search engine should get results.)

C.2 Gramm–Schmidt Orthonormalisation Process

Even in an applied text such as this, it is important that we know formally how to construct an orthonormal set of basis vectors from a given basis. The Gramm–Schmidt process gives an infallible method of doing this. We state this process in the form of a theorem and prove it.

Theorem C.20 *Every finite dimensional inner product space has a basis consisting of orthonormal vectors.*

Proof Let $\{v_1, v_2, v_3, \ldots, v_n\}$ be a basis for the inner product space V. A second equally valid basis can be constructed from this basis as follows

$$u_1 = v_1$$

$$u_2 = v_2 - \frac{(v_2, u_1)}{\|u_1\|^2} u_1$$

$$u_3 = v_3 - \frac{(v_3, u_2)}{\|u_2\|^2} u_2 - \frac{(v_3, u_1)}{\|u_1\|^2} u_1$$

$$\vdots$$

$$u_n = v_n - \frac{(v_n, u_{n-1})}{\|u_{n-1}\|^2} u_{n-1} - \cdots - \frac{(v_n, u_1)}{\|u_1\|^2} u_1$$

where $u_k \neq 0$ for all $k = 1, 2, \ldots, n$. If this has not been seen before, it may seem a cumbersome and rather odd construction; however, for every member of the new set $\{u_1, u_2, u_3, \ldots, u_n\}$ the terms consist of the corresponding member of the start basis $\{v_1, v_2, v_3, \ldots, v_n\}$ from which has been subtracted a series of terms. The coefficient of u_j in u_i $j < i$ is the inner product of v_i with respect to u_j divided by the length of u_j. The proof that the set $\{u_1, u_2, u_3, \ldots, u_n\}$ is orthogonal follows the standard induction method. It is so straightforward that it is left for the reader to complete. We now need two further steps. First, we show that $\{u_1, u_2, u_3, \ldots, u_n\}$ is a linearly independent set. Consider the linear combination

$$\alpha_1 u_1 + \alpha_2 u_2 + \cdots + \alpha_n u_n = 0$$

and take the inner product of this with the vector u_k to give the equation

$$\sum_{j=1}^{n} \alpha_j (u_j, u_k) = 0$$

from which we must have

$$\alpha_k(u_k, u_k) = 0$$

so that $\alpha_k = 0$ for all k. This establishes linear independence. Now the set $\{w_1, w_2, w_3, \ldots, w_n\}$ where

$$w_k = \frac{u_k}{\|u_k\|}$$

is at the same time, linearly independent, orthogonal and each element is of unit length. It is therefore the required orthonormal set of basis vectors for the inner product space. The proof is therefore complete.

□

Bibliography

Bolton, W. (1994) *Laplace and z-transforms* 128pp., Longmans.

Bracewell, R.N. (1986) *The Fourier Transform and its Applications (2nd Edition)* 474pp., McGraw-Hill.

Churchill, R.V. (1958) *Operational Mathematics* 337pp., McGraw-Hill.

Copson, E.T. (1967) *Asymptotic Expansions* 120pp., C.U.P.

Hochstadt, H. (1989) *Integral Equations* 282pp., Wiley-Interscience.

Jones, D.S. (1966) *Generalised Functions* 482pp., McGraw-Hill. (new edition 1982, C.U.P.)

Lighthill, M.J. (1970) *Fourier Analysis and Generalised Functions* 79pp., C.U.P.

Needham, T. (1997) *Visual Complex Analysis* 592pp., Clarendon Press, Oxford.

Osborne, A.D. (1999) *Complex Variables and their Applications* 454pp., Addison-Wesley.

Pinkus, A and S. Zafrany (1997) *Fourier Series and Integral Transforms* 189pp., C.U.P.

Priestly, H.A. (1985) *Introduction to Complex Analysis* 157pp., Clarendon Press, Oxford.

Sneddon, I.N. (1957) *Elements of Partial Differential Equations* 327pp., McGraw-Hill.

Spiegel, M.R. (1965) *Laplace Transforms, Theory and Problems* 261pp., Schaum publ. co.

Stewart, I. and D. Tall (1983) *Complex Analysis* 290pp., C.U.P.

Watson, E.J. (1981) *Laplace Transforms and Applications* 205pp., Van Nostrand Rheingold.

Weinberger H.F. (1965) *A First Course in Partial Differential Equations* 446pp., John Wiley.

Whitelaw, T.A. (1983) *An Introduction to Linear Algebra* 166pp., Blackie.

Williams, W.E. (1980) *Partial Differential Equations* 357pp., O.U.P.

Zauderer, E. (1989) *Partial Differential Equations of Applied Mathematics* 891pp., John Wiley.

Index

absolutely convergent, 24
amplitude, 128
analytic functions, 156
arc, 158
Argand diagram, 155
asymptotics, 122, 175, 179
autocovariance function, 150

basis, 234
bending beams, 68
Bessel Functions, 180
Bessel's inequality, 77
Borel's theorem, 39
boundary condition, 120
boundary conditions, 142
boundary value problem, 121, 141
branch point, 158
branch points, 165, 175
Bromwich contour, 168, 171, 173

cantilever, 119
capacitor, 59
Cauchy's integral formulae, 157
Cauchy's theorem, 159
Cauchy–Riemann equations, 156, 160
Cauchy–Schwartz inequality, 237
celerity, 112, 119
chain rule, 110
characteristics, 112
complementary error function, 45
complex analysis, 155

complex variable, 155
complex variables, 122
complimentary error function, 174
contour, 159
contour
    closed, 159
    simple, 159
contour integral, 159
contour integrals, 163
convergence, 115
convergence in the norm, 242
convergence, pointwise, 78
convergence, uniform, 78
convolution, 37, 73, 144
convolution
    for Fourier Transforms, 144
convolution theorem, 38, 48, 54, 144
cut (complex plane), 166

diffusion, 45
Dirac's delta function, 26, 125, 128,
    132
Dirac's delta function
    derivative of, 32
Dirac,s delta function
    derivatives of, 30
Dirichlet's theorem, 79
discontinuous functions, 9

electrical circuits, 55
energy spectral density, 148